ENERGY RESOURCES

Geology, Supply and Demand

G.C. BROWN *and* E. SKIPSEY

The Open University

Open University Press

Milton Keynes · Philadelphia

Open University Press
Open University Educational Enterprises Limited
12 Cofferidge Close
Stony Stratford
Milton Keynes MK11 1BY, England

and
242 Cherry Street
Philadelphia, PA 19106, USA

First Published 1986

British Library Cataloguing in Publication Data

Energy resources : geology, supply and demand.
   1. Power resources
   I. Brown, G.C.   II. Skipsey, E.
   333.79   TJ163.2

   ISBN 0 335 15200 7

Library of Congress Cataloging in Publication Data
Main entry under title:
Brown, G. C. (Geoff C.)
   Energy resources.

   Bibliography: p.
   Includes index.
   1. Energy minerals. 2. Energy minerals — Great Britain. I.
Skipsey, E. (Eric) II. Title.
TN263.5.B76 1986   333.79'11   86 12596

   ISBN 0 335 15200 7

Printed in Great Britain

Contents

Preface   vi
Acknowledgements   vii

PART I

1   Introduction to energy resources   **1**
1.1   Primary energy resources   2
1.1.1   External energy sources
1.1.2   The Earth's internal heat
1.2   Secondary energy resources   6
1.2.1   Photosynthesis and fossil fuels
1.2.2   The fossil fuel bank
1.3   Summary of Chapter 1   7

2   Energy resources: some global perspectives   **9**
2.1   The growth of energy demand: a historical perspective   9
2.2   Global energy requirement: a numerical perspective   11
2.3   Energy resource distribution: a geological perspective   13
2.4   Summary of Chapter 2   15

3   Geology of coal deposits   **16**
3.1   Origins of coal   16
3.2   Sedimentary environment of coal deposits   17
3.3   Coal formation   19
3.3.1   Causes of rank change in coal
3.4   Properties of coal   22
3.4.1   Impurities in coal
3.5   Ages of coal deposits   24
3.6   Summary of Chapter 3   27

4   Coal: exploration and extraction   **29**
4.1   Locations of the coalfields in Britain   29
4.2   Exploration and evaluation techniques   32
4.2.1   Drilling
4.2.2   Seismic reflection profiling
4.2.3   Exploration in Britain
4.3   Surface mining   39
4.4   Underground mining   40
4.4.1   Structural setting
4.4.2   Sedimentary setting
4.4.3   Recognition of geological hazards
4.4.4   Coal extraction
4.4.5   Processing the coal
4.5   Environmental aspects of coal mining   45
4.6   Uses of coal   46
4.6.1   Future trends in coal use
4.7   British coal resources   49
4.8   Summary of Chapter 4   50

5   The world coal economy   **51**
5.1   International trade in coal   51
5.2   World coal resources   53
5.3   Summary of Chapter 5   54

6 The nature, origin and generation of petroleum   **55**
6.1 Nature of petroleum   55
6.2 Origin of petroleum   55
6.3 Generation of petroleum   57
6.4 Summary of Chapter 6   62

7 Migration and accumulation of petroleum   **63**
7.1 Sedimentary basins   63
7.2 Migration of petroleum   63
7.3 Reservoirs   64
7.4 Traps   64
7.4.1 Structural traps
7.4.2 Stratigraphic traps
7.4.3 Combination traps
7.5 Natural gas   69
7.6 Solid petroleum and oil shale   70
7.7 Summary of Chapter 7   71

8 Petroleum exploration, evaluation and recovery techniques   **72**
8.1 Exploration techniques   72
8.2 Trap structure and the recovery of oil   72
8.3 Primary, secondary and enhanced recovery methods   74
8.4 Summary of Chapter 8   75

9 Petroleum in Britain, and its world setting   **76**
9.1 Onshore exploration in Britain   76
9.2 Onshore to offshore   76
9.2.1 North Sea petroleum
9.2.2 Structure of the North Sea basin
9.3 Middle East basin   84
9.4 World resources of petroleum   87
9.5 Output and consumption   88
9.6 Coal versus petroleum   90
9.7 Summary of Chapter 9   91

Further reading for Part I   92

PART II

1 Nuclear power   **95**
1.1 Nuclear reactions and reactors   95
1.1.1 Types of nuclear reactor and fuel requirements
1.1.2 Nuclear power in Britain
1.2 Uranium geology   105
1.2.1 The geochemistry of uranium
1.2.2 Pegmatites and disseminated magmatic deposits
1.2.3 Vein-type deposits
1.2.4 Uranium in sandstones
1.2.5 Quartz-pebble conglomerate deposits
1.2.6 Other sedimentary uranium deposits
1.3 Uranium production and economics   117
1.3.1 Mining and milling of uranium
1.3.2 Uranium reserves and resources
1.4 Summary of Chapter 1   128

2	**Geothermal energy**	**130**
2.1	Hyper-thermal resources	132
2.2	Zones with shallow, low conductivity strata	137
2.3	Hot dry rocks	140
2.4	Summary of Chapter 2	143
3	**Surface energy resources**	**144**
3.1	Solar energy	144
3.1.1	Thermal collection of solar energy	
3.1.2	Photovoltaic conversion of solar energy	
3.1.3	Energy from the biomass via photosynthesis	
3.1.4	Solar energy: a UK perspective	
3.2	Wind energy	155
3.3	Hydroelectric power	158
3.3.1	Contribution of hydropower to global power supplies	
3.4	Tidal power	162
3.5	Wave energy	166
3.6	Summary of Chapter 3	168
4	**Side-effects of energy conversion**	**170**
4.1	Side-effects of fossil fuel and biomass conversion	171
4.2	Side-effects of the nuclear power industry	175
4.2.1	Radioactive waste disposal	
4.3	Excess heat	178
4.4	Summary of Chapter 4	180
5	**The future of energy supply and demand**	**181**
5.1	Energy supply, end-uses and conservation in the UK	181
5.2	The future world energy scene — prospects and possibilities	186
5.3	Summary of Chapter 5	191
	Further reading for Part II	192
	Self-assessment questions for Part I	193
	Self-assessment questions for Part II	198
	SAQ answers and comments for Part I	201
	SAQ answers and comments for Part II	205
	References	209
	Appendix	210
	Index	211

Preface

Energy resources are of fundamental importance to our way of life and civilization. Consequently their availability, their geological and geographical distribution around the world and the economic and political power that control of their supply can provide have made the study of energy resources an essential element in our understanding of the modern world.

This book has been prepared as an edited version of the energy resource components taken from the Open University earth sciences course S238 *The Earth's Physical Resources*. The book is designed to provide an understanding of the following aspects of the subject:

1. The main energy resources and their uses
2. The geological processes leading to their formation
3. Their global distribution and the quantities of energy reserves
4. The methods used to locate recoverable reserves and to extract them from the ground
5. The factors controlling the supply and demand of energy resources
6. Some of the environmental and social implications resulting from the exploitation of the various energy resources

The book has been prepared in two parts, with the first part dealing with those energy resources generally classed as fossil fuels. Chapters 1 and 2 set the scene for the whole book with an introduction which considers the types of energy resources and the global demand for energy. Chapters 3 to 9 then deal with the formation, geological distribution, extraction and uses of coal, petroleum and natural gas.

Part II of the book focusses on the non-hydrocarbon energy resources with studies of nuclear and geothermal energy forming the first two chapters. The renewable or alternative energy resources are then reviewed in Chapter 3. Costs of energy exploitation in terms of atmospheric pollution and other side effects, including the problem of the safe disposal of nuclear wastes, are considered in Chapter 4. Finally Chapter 5 summarises energy supply and demand in the light of changing energy supply costs and of increasing energy conservation measures.

Acknowledgements

This text is up-dated and based on an Open University text prepared by the S238 (Earth's Physical Resources) Course Team with academic authorship as follows:–

Geoff Brown
Steve Drury
Ian Gass (Course Team Chair)
Graham Jenkins
Patricia McCurry (Consultant)
Isla McTaggart (Course Coordinator)
Dave Park
Julian Pearce
Eric Skipsey
Peter Smith
Sandra Smith
John Wright

The present editors are grateful to Dick Sharp who was the original editor for much of this text, to Janis Gilbert and Pam Owen (illustrators) and Jane Sheppard (Designer). They also wish to thank Ann Budd and Jean Elsy for secretarial services.

Grateful acknowledgement is made to the following sources for figures used in Part I:

Figure 4 adapted from D. C. Ion 'World oil and gas' in *Report No. 9 Assessment of Energy Resources,* The Watt Committee on Energy Ltd, 1981; *Figure 5* from A. J. Smith 'Geological constraints on conventional energy resources' in *Report No. 9 Assessment of Energy Resources,* The Watt Committee on Energy Ltd, 1981; *Figure 6* The National Coal Board; *Figure 10* from G. C. Gester in B. J. Skinner, *Earth Resources,* Prentice-Hall Inc. 1976 (2nd edn); *Figure 12* based on a map by Prof. A. J. Smith; *Figure 13* from S236 Block 6 Fig. 3 (pair). *Figure 14* from K. Moses, 'Britain's coal resources and reserves' in *Report No. 9 Assessment of Energy Resources,* The Watt Committee on Energy Ltd; *Figure 15* from R. Anderton *et al., A Dynamic Stratigraphy of the British Isles,* Allen & Unwin, 1979; *Figure 21* from W. E. Lerwill and A. M. Ziolkowski, 'A simple approach to high resolution seismic profiling for coal' in *Geophysical Prospecting,* vol. 27, 1979, European Association of Exploration Geophysicists; *Figure 22* from R. Goosens, Coal Exploration in Britain, *Colliery Guardian,* August 1985; *Figure 23b* from the National Coal Board, Opencast Executive. *Figure 27* from Commission on Energy and the Environment, *Coal and the Environment,* 1981, reproduced by permission of the Controller of HMSO; *Figure 28* reprinted by permission from *Nature,* vol. 277, pp. 463–465, copyright © 1979, Macmillan Journals Ltd; *Figures 46, 48 and 50,* from D. H. Tarling (ed.) *Economic Geology and Geotectonics,* Blackwell Scientific, 1981; *Figure 32* from J. M. Hunt, *Geochemistry of Petroleum,* American Association of Petroleum Geologists; *Figures 33(a) and (b)* from B. P. Tissot and D. H. Welte, *Petroleum Formation and Occurrence,* Springer-Verlag Inc., 1978; *Figure 47* from R. Stoneley *et al., Introduction to Petroleum Geology for Non-Geologists,* JAPEC (UK) c/o Geological Society of London; *Figure 45* from A. J. Stauble and G. Milius in M.T. Halbouty (ed.) *Geology of Groningen Gas Field, Netherlands,* Memoir no. 14, 1970, American Association of Petroleum Geologists.

The Course Team acknowledges with gratitude the assessors, Professor P. McL. D. Duff (BP Coal Ltd) and Professor R. Stoneley (Imperial College, London), for their constructive criticism of the text, and Dr E. L. Boardman, Dr S. Brassell, Dr G. T. George, Dr T. B. H. Jenkins, J. Slater, M.D. Wright and officers of the National Coal Board, for their advice.

Grateful acknowledgement is made to the following sources for material used in Part II:

Figures 1 and 4 COI, *Nuclear Energy in Britain,* Pamphlet no. 28, reproduced by permission of the Controller of HMSO; *Figure 5* courtesy of UKAEA; *Figures 8 and 12* based on IUREP, *World Uranium — Geology and Resource Potential,* OECD, 1980; *Figures 22–24* International Atomic Energy Agency/OECD, *Uranium–Resources, Production and Demand,* OECD, 1983; *Figure 26* G. W. Grindley, *The Geology, Structure, and Exploitation of the Wairakei Geothermal Field, Taupo, New Zealand,* New Zealand Geological Survey, 1965; *Figures 32 and 49* G. Long et al., *Solar Energy,* Department of Energy Paper no. 16, 1976, reproduced by permission of the Controller of HMSO; *Figure 34* B. W. Hatt and A. V. Bridgewater, *Energy from the Biomass,* Watt Committee on Energy Report no. 5, 1979; *Figures 35, 41(b) and 54* D. Crabbe and R. McBride, *The World Energy Book,* Kogan Page Ltd., 1978; *Figures 40, 41(a) and 50 Tidal Power from The Severn Estuary,* vol. 1, Department of Energy Paper no. 46, 1981, reproduced by permission of the Controller of HMSO; *Figure 42 Wave Energy,* Department of Energy Paper no.42, 1979, reproduced by permission of the Controller of HMSO; *Figure 43* National Engineering Laboratory, UK Crown copyright; *Figure 45* OECD, 1977; *Figure 46* K. D. B. Johnson, The disposal of high-level radioactive wastes, in The British Association Meeting 1979 paper *Energy in the Balance,* Butterworths; *Figure 48(a)* G. Wick, Proposed salt repository for nuclear wastes, in *New Scientist,* 16 March 1972; *Figure 52 Energy Technologies for UK,* vol. 1, Department of Energy Paper no.39, 1979, reproduced by permission of the Controller of HMSO; *Figure 53* C. M. Summers, The conversion of energy, in *Scientific American,* Sept. 1971, © 1971 by Scientific American, Inc. All rights reserved.

The editors acknowledge with gratitude the following Open University assessors for their constructive criticism of parts of the draft text: Dr G. Long (Energy Technology Support Unit, Harwell) and Dr R. C. Malan (US Department of Energy, Grand Junction, Colorado).

PART I

1 Introduction to energy resources

Study Comment. We begin by reviewing and classifying the various types of energy resource and then describe the internal and external energy sources that affect the Earth. This Chapter then examines the link between photosynthesis and fossil fuels.

In this book, you will be studying resources which, in commercial and industrial terms, are the most fundamental of all natural raw materials used by modern society. Almost all other resources depend on *energy* for their extraction and processing; moreover, the availability and cost of energy often exerts the predominant control on the economics of metals, bulk materials and, to a lesser extent, water. If, in the future, a cheap and plentiful supply of energy could be made available, then currently uneconomic grades of ore, depths of mining, bulk materials processing and water purification technologies might all become more economically feasible.

To appraise the available options, a full understanding of each of the different energy resources is needed, including their origin, occurrence, extraction, processing and availability.

It is convenient to subdivide the available energy resources into primary and secondary categories. *Primary energy resources** are those affecting the Earth in a natural way (such as solar energy) which may be harnessed for human exploitation. *Secondary energy resources** are those in which primary energy has been converted into a stored form (coal, for example) suitable for extraction and reconversion into electrical and other forms of usable energy.

Which of these two categories of energy provide *renewable* and which *non-renewable resources* on the scale of human lifetimes?

You should realise that primary energy resources, such as the heat of the Sun, are renewable and are sometimes termed *energy income*, whilst secondary energy resources, such as coal, are being exploited much more rapidly than they are formed and so are non-renewable; they are *energy capital*.

You should note that several different units are in common use for energy reserves or energy demand. Units such as tonnages of coal, barrels of oil, cubic metres of gas, etc., can all be reduced to a common currency — the SI unit of energy *joules*. Since *power* is the rate of energy supply obtainable from primary energy resources it can be compared with the total number of joules in the various 'banks' of energy capital. To give you an impression of how valuable a joule of energy is, a power supply of 60 joules per second, that is 60 watts, will run a 60 watt light bulb, whereas a supply of 1000 joules per second, that is 1000 watt, or 1 kilowatt, will run a one-bar electric fire. After 1 hour, a power supply of 1 kilowatt will have used 1 *kilowatt hour* (kWh) of energy (i.e. 1000×3600 (60 seconds $\times$ 60 minutes) $= 3.6 \times 10^6$ joules): over three and a half million joules are equivalent to 1 kWh, or a single 'unit' on domestic electricity bills. All these forms of energy and power are summarized and conversion factors are given in the data supplements.

* **You may come across the following different usage of the terms:** *primary energy resources* for fuels such as coal, petroleum and uranium; *secondary energy resources* for the marketed products such as coke, petrol and electricity. Our definitions in the context of Earth Science are more appropriate to this book.

1.1 Primary energy resources

Primary sources of energy can be classified into two main kinds:

A *External energy sources*
   1 Solar and other cosmic radiation
   2 Gravitational influence of the Sun and the Moon

B *Internal energy sources*
   1 Earth's internal heat
   2 Earth's rotation and gravity field

1.1.1 External energy sources

You will realise that the external source of energy that has most effect on natural processes is solar radiation. This is responsible for the atmospheric pressure differences that cause winds (and hence ocean waves), evaporation and reprecipitation of water vapour in the *hydrological cycle* , and *photosynthetic activities* among organisms. These solar-induced effects have controlled the change of surface environments in which life on Earth originated, evolved and became fossilized within sedimentary rock sequences. The origin, potential use and ultimate fate of energy from both external sources are summarized in Figure 1, which you should study carefully, noting in particular the following points.

1 The solar power that is actually absorbed by the surface and in the atmosphere (3.5×10^{24} J yr^{-1}) is only about 65 per cent of the amount that reaches the Earth from the Sun (5.4×10^{24} J yr^{-1}); the rest is reflected back into space.

2 The Earth's natural power supply is made up of:

5.4×10^{24} J yr^{-1} from the Sun

ca. 10^{20} J yr^{-1} from gravitational energy

ca. 10^{21} J yr^{-1} from internal energy sources

In other words, the Sun's contribution to this total exceeds 99.9 per cent.

Solar radiation

In common with all stars, the Sun is a vast nuclear powerhouse giving clues to the ultimate origin of most forms of energy and even of the chemical elements from which stars and planetary systems are made. Its thermal energy is released by *nuclear fusion* reactions. On an annual basis the total solar power output is 1.2×10^{34} J but, of course, only a tiny fraction of this, some 5.4×10^{24} J, is intercepted by the Earth and only 65 per cent of this (3.5×10^{24} J yr^{-1}) is absorbed in the atmosphere or at the Earth's surface (cf. Figure 1). The Earth's curvature causes the heating effect to be greater in equatorial latitudes than at the poles, and this differential heating combines with the Earth's rotation to produce parallel belts of high and low atmospheric pressure between which winds blow (Figure 2). About 10 per cent of the absorbed solar power (*ca.* 3×10^{23} J yr^{-1}) gives rise to winds. Ultimately, much of the kinetic energy of winds is dissipated as heat; however, some 10 per cent of 'wind energy' (*ca.* 3×10^{22} J yr^{-1}) is converted first into water waves through frictional effects at the sea surface. Although there are huge practical and technical problems to be overcome before wind power and wave power can be harnessed in any major way, there is no shortage of ideas and some successes have been recorded; these are discussed in Part II.

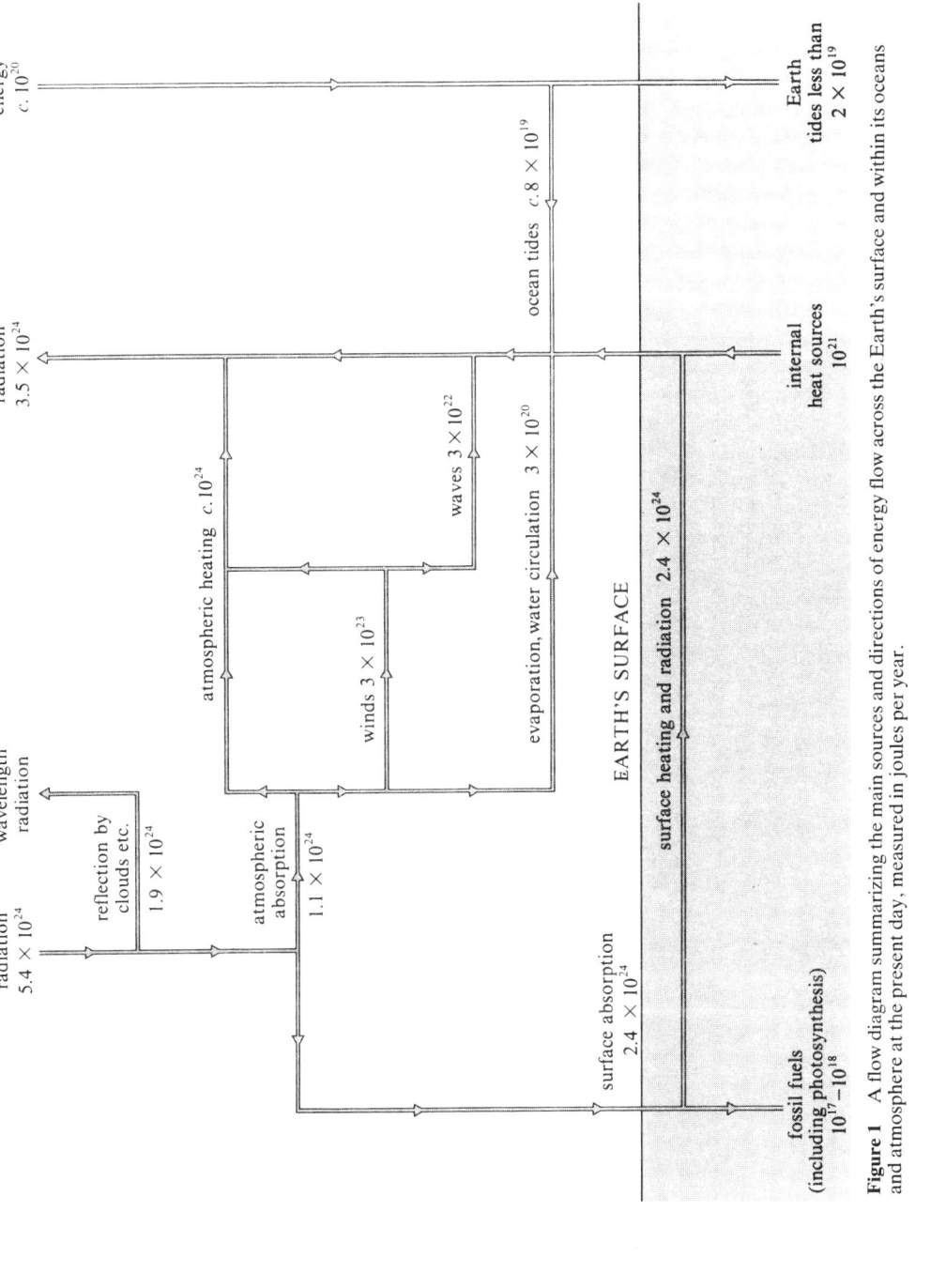

Figure 1 A flow diagram summarizing the main sources and directions of energy flow across the Earth's surface and within its oceans and atmosphere at the present day, measured in joules per year.

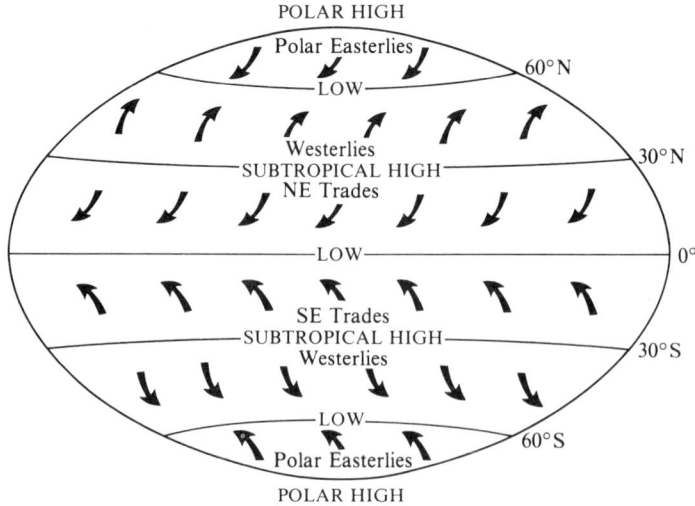

POLAR HIGH

Figure 2 The main belts of high and low pressure together with the associated wind systems (arrows) caused by the combined effects of differential solar heating and the Earth's rotation. This diagram assumes that the Sun is overhead at the equator (as at the Spring and Autumn equinoxes).

Another small portion (*ca.* 3×10^{20} J yr^{-1}) of the absorbed solar power appears as gravitational energy in the hydrological cycle, when water evaporated from seas, lakes and so on, returns to the surface as precipitation. This gravitational energy is normally dissipated as heat during the return flow of the water to the oceans, but some may be harnessed to provide hydroelectric power, discussed in Part II.

On reaching the Earth's surface, solar radiation encounters surface water, vegetation cover, rocks or soil. A small proportion of this energy is potentially available for consumption as an energy resource either by direct conversion of solar energy or via biological production and conversion. About half the energy input of 2.4×10^{24} J yr^{-1} is used to evaporate surface water and is thereby transferred back to the atmosphere whilst the other half is used to heat the surface. Of course, most of this surface energy also is ultimately re-radiated into the atmosphere. However, the atmosphere is a complex system which, because it contains carbon dioxide and water vapour, is more 'transparent' to incoming short-wavelength solar radiation than it is to outgoing long-wavelength terrestrial radiation. The only way that an equilibrium can be established where-by the Earth's heat gain and losses from the upper atmosphere are balanced is for the temperature of the atmosphere to remain at a high mean value of 15°C. If on the other hand the atmosphere were transparent to terrestrial radiation then the equilibrium temperature would be an inhospitable −18°C. The raising of global temperatures due to minor atmospheric constituents is known as the *greenhouse effect*.

Table 1 completes this introduction to solar radiation by providing both a balance sheet of solar power expenditure and also data on the potential availability of various forms of energy.

Gravitational influence of the Sun and the Moon

It is impossible to separate discussion of the Earth's axial rotation and gravity field, defined in Section 1.1 as an internal energy source, from that of the external gravitational influence of the Sun and the Moon, because the two effects are both necessary to produce *tidal power*, which is exploitable as an energy resource. The amount of power associated with tides can be calculated from a knowledge of the rotational energy of the Earth as about 10^{20} J yr^{-1}, many orders of magnitude less than the solar input (see Figure 1). About 8×10^{19} J yr^{-1}, appear as the kinetic energy of water motion in the oceans and are converted ultimately into heat, mainly through friction between water and land; the remaining 2×10^{19} J yr^{-1} take the form of solid Earth tides that are of no real consequence.

Table 1 Summary of solar power expenditure and its likely maximum potential as an energy resource

Form of energy expenditure[a]	Total energy involved/ J yr^{-1}	Maximum potential usage/ J yr^{-1}
direct reflection	1.9×10^{24}	
direct atmospheric heating (includes dissipation of wind energy)	1.1×10^{24}	
atmospheric circulation (winds and waves)	3.0×10^{23}	*c.* 1.0×10^{18}
gravitational energy of reprecipitated water	3.0×10^{20}	*c.* 1.0×10^{20}
surface heating { water evaporation[b]	1.2×10^{24}	
heating and long wavelength radiation[b]	1.2×10^{24}	*c.* 2.4×10^{20}
biological fuel (via photosynthesis)[c]	1.0×10^{22}	*c.* 1.0×10^{20}

Notes on Table 1

(*a*) The total solar power income (5.4×10^{24}J yr^{-1}) is balanced by three main forms of energy expenditure: direct reflection, atmospheric heating and surface heating ($1.9 + 1.1 + 2.4 = 5.4$).

(*b*) All the surface heating component is ultimately transferred back into space by direct radiation or via the clouds and atmosphere; atmospheric heating contributes to long-wavelength radiation as well.

(*c*) About 10^{22} J yr^{-1} of solar energy is converted into plant material but the input to fossil fuels amounts to only $10^{17}–10^{18}$ J yr^{-1}; the reason for this difference is discussed in Section 1.2.

Some of the energy of ocean tides can be exploited along coastlines where the natural tidal range (*ca.* 1 metre) is accentuated by geographical factors (e.g. tidal amplitudes of 12 metres are recorded off the coast of NE Canada). Techniques for exploiting tidal power are discussed in Part II.

1.1.2 The Earth's internal heat

The occurrence of both volcanoes and hot springs is a graphic illustration that the Earth's interior is hot: it can produce molten rock at temperatures of up to 1250°C and superheated steam. However, these phenomena are mainly confined to several narrow and elongate zones along *active continental margins* and *ocean-ridge* zones, the currently active boundaries of the *tectonic plates*. Figure 1 shows that the power output from the Earth's interior, *ca.* 10^{21} J yr^{-1}, is about 5000 times smaller than the available solar power. Many measurements have now been made of the surface heat flow from the Earth's interior and, except for the distinctive volcanic zones just considered, this heat flow is remarkably constant, averaging 60 milliwatts (mW) per square metre over the whole Earth. It is possible to exploit this *geothermal heat* in areas of recently active volcanoes; this is not the only possibility as we will explain in Part II.

What is the source of the Earth's internal heat?

The heat energy comes from the decay of long-lived *radioactive isotopes* . The most important of these are potassium-40, thorium-232, uranium-235 and uranium-238. Of these ^{235}U, which comprises only 0.7 per cent of natural uranium, forms the basis of another major category of energy resources: *nuclear fuels* (discussed in Part II).

1.2 Secondary energy resources

The *fossil fuels*, coal, oil and natural gas, are secondary energy resources because they are all forms of stored solar energy (uranium used to fuel nuclear reactors is also a secondary resource, for it represents primary nuclear energy stored when the Earth first formed.)

1.2.1 Photosynthesis and fossil fuels

All living things on the Earth's surface depend on the trapping of solar radiation by plants. This is true even of higher vertebrates, which eat either plants or animals that in turn have eaten plants. The processes by which land plants and *phytoplankton* in the oceans convert carbon dioxide and water into carbohydrate and oxygen is known as photosynthesis. It can be summarized by a simple *endothermic reaction*:

$$6CO_2 + 6H_2O + \underset{\text{solar energy}}{2.8 \times 10^6 \text{ J}} = \underset{\text{carbohydrate}}{C_6H_{12}O_6} + 6O_2 \qquad (1)$$

The resulting carbohydrate stores energy until it is broken down by the reverse *exothermic reaction*, when it gives off heat during combustion, or decay caused by metabolic activity or by some other means.

About 1×10^{22} J yr^{-1} (*ca.* 0.2 per cent) of the Earth's incident solar power (cf. Table 1) is converted into plants, which together with animals that feed on them are known as the *biomass*. Photosynthesis continuously creates a biomass with an energy content some thirty times greater than the total energy demand of humankind (3×10^{20} J yr^{-1}), but there are many difficulties associated with converting enough of the fuel to supply our energy needs.

In an ideal biological cycle, all organic matter is ultimately consumed by decay or combustion so that the intake and output of energy expressed by reaction 1 and its reverse are exactly balanced. The heat balance of the biosphere and of the Earth's surface itself would be disturbed if there were departures from this equilibrium. Over the thousands of millions of years since life appeared on Earth, there have in fact been minor departures from this equilibrium. Some organic matter has accumulated in oxygen-free (*anaerobic* or anoxic) conditions, such as swamps, bogs and parts of the ocean floor, and this material has not decayed. The existence of fossil fuels, which are the products of undecayed organic matter, means that there is undecayed organic matter trapped in sedimentary rocks. We shall later examine some of the chemical reactions that have turned these remains into valuable fossil fuels.

1.2.2 The fossil fuel bank

The fossil fuel bank that has accumulated over the past few hundred million years in particular (the period of major coal and petroleum formation) represents humankind's most important energy resource to date. It contains some 10^{23} joules and is being added to at the rate of 10^{17}–10^{18} J yr^{-1} (Figure 1). But we must emphasize that this represents less than a thousandth part (in fact, it is about 1/2500 part, or 0.04 per cent) of the total amount of undecayed or fossilized organic *hydrocarbons*. Most of the organic material occurs in sediments and sedimentary rocks as finely dispersed animal and plant debris that will never be commercially exploitable. The *average* carbon content of sediments and sedimentary rocks is about 0.4 per cent. Despite their apparently widespread occurrence (See Figures 12 & 49), coalfields and oil fields represent a limited number of locations where this average has been greatly exceeded as a result of geological processes.

1.3 Summary of Chapter 1

1   Primary energy resources represent a continuous, renewable power input, or energy income, to the Earth's energy budget whereas secondary energy resources are non-renewable forms of energy capital. There are internal and external forms of both primary and secondary energy resources.

2   The Sun is by far the most important primary source of energy. Of the solar radiation reaching the Earth, 35 per cent is reflected and 65 per cent is absorbed by the Earth's atmosphere and surface, contributing to winds, waves, atmospheric water circulation, atmospheric heating, surface water evaporation, etc. (Figure 1).

3   The external gravitational influence of the Sun and the Moon combined with the Earth's axial rotation produces tidal effects in the oceans, which are potentially exploitable sources of power.

4   Various minor constituents in the atmosphere render it less transparent to outgoing terrestrial radiation than to incoming solar radiation. This results in a global average of 15 °C for the equilibrium temperature of the surface.

5   A small contribution to the surface energy balance is made by radioactive decay from long-lived isotopes of potassium, thorium and uranium. Although the surface heat flow from the Earth's interior is only 1/5000 of that provided by the Sun, nevertheless at depths within the Earth beyond the influence of solar radiation (i.e. greater than a few metres), geothermal energy is most important.

6   Secondary energy resources include uranium reserves and fossil fuels that can be extracted from the Earth's crust. Fossil fuels represent a minute fraction

of the solar power that has reached the Earth during the past 400 Ma (the time since land plants first appeared) and has become converted through photosynthesis of plants and anaerobic decomposition, into hydrocarbon form. Only a small fraction of the hydrocarbons so produced is sufficiently concentrated to be commercially extractable as secondary energy resources; these comprise the fossil fuel bank.

2 Energy resources: some global perspectives

Study Comment. Subsequent Chapters of this book are devoted to a detailed examination of individual energy resources and so, to complete the basic framework on which these sections build, we develop here three aspects in a global perspective. First we consider the historical development of energy use, then we give some estimates of the likely future demand for energy. Finally, we provide an outline of the geological distribution of energy resources.

2.1 The growth of energy demand: a historical perspective

Human beings have always depended on energy to grow and prepare their food, to heat buildings, and in recent centuries, to manufacture, power and maintain increasingly sophisticated machinery. Generally, the energy source adopted has been the most convenient in terms of location and use. Thus, before the Industrial Revolution, easily harnessed energy came from the burning of wood, and from wheels and pumps driven by wind or water. The use of these primitive sources has far from disappeared; for example, approximately one-sixth of the amount of the world's annual fuel supplies are wood fuel and roughly half the trees cut down are used for cooking and heating purposes. About half the world's population still relies mainly on wood for domestic fuel and one recent (probably conservative) estimate is that about 10^9 tones of wood are burned each year, equivalent to a power output of 2×10^{19} J yr^{-1}, or 7 per cent of the energy consumed through commercial outlets in the form of electricity, gas, etc.

The use of coal in Western Europe as a source of domestic heating can be traced back through the Middle Ages to the twelfth century. However, it was not until about AD 1500 that coal was widely transported for use as an energy resource; exhaustion of local wood supplies was the prime cause for the change from wood to coal.

> Coal had two important advantages over wood as an energy resource in the Middle Ages. Can you suggest what they were?

Coal is a more concentrated form of energy than wood and so, for an equivalent energy output, it was cheaper to transport. It also produces a hotter flame than wood, which made the working of iron by blacksmiths easier. Production from British coalfields is estimated to have been about 200 000 tonnes a year by 1600, increasing rapidly during the Industrial Revolution to reach an all-time peak of 285 million tonnes a year by 1913, following the great Victorian age of increased mechanization, proliferation of the railways and the introduction of electricity distribution networks. Of course, industrialization came early in the Western world and, even in the USA, energy production from coal surpassed that from wood as early as 1885 (Figure 3). But the growth of manufacturing industry on a world scale has continued to the present day and, with it, world coal production has continued to increase.

Perhaps the most important development ever to affect the world energy scene came in the early years of this century with the invention of assembly line car manufacturing (attributed to Henry Ford in 1913). This has brought private car ownership to almost one third of the world's population and has created an enormous demand for petroleum products. Most nations have shown a rapid increase in their consumption of oil and gas since 1920; the USA (Figure 3) led the world in consuming more oil than coal, in terms of energy equivalent, by 1950 (1966 in Western Europe), and in reaching the peak of production from its own oil reserves by 1970. The period from 1950 to 1970 saw immense changes in the pattern of world energy supply and demand. Not only were most developed nations changing from being just major coal consumers to being major oil, gas and coal consumers, but many of them started to rely heavily on imports of oil and gas from foreign sources, particularly from the Middle East where 55 per cent of the world's oil reserves are located (27 per cent in Saudi Arabia alone).

This had several important and irreversible political and economic implications. International co-ordinating bodies concerned with energy policy and supply came into being. The OPEC (Organization of Petroleum Exporting Countries)

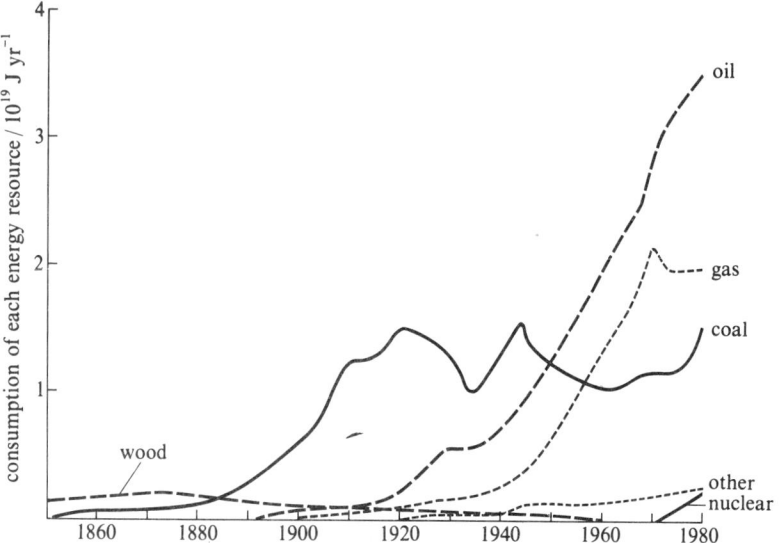

Figure 3 The main sources of power used in the USA from 1850 to 1980

cartel was formed by thirteen oil-producing countries in 1960 with the aim of controlling production according to reserves. In the early 1980s OPEC controlled 70 per cent of the proved reserves outside the communist world, notably in the Middle east, North Africa, Nigeria and Venezuela. There were problems, however, both in the determination of price structures and also because some of the member states, such as Venezuela, Libya and Kuwait, preferred to restrict productions in order to conserve their reserves. The cartel has found it difficult to agree production totals for its members, and oil prices were often determined by the dominant member, Saudi Arabia. Although the Saudis provided a voice of moderation in the OPEC discussions of oil prices, the market price of crude oil rose from 3 US dollars a barrel in 1973 to 35 US dollars a barrel in 1982 whilst production costs remained almost constant!

The volatility of oil prices is highlighted by the dramatic fall in 1986 to around 10 US$ per barrel and by further attempts by O.P.E.C. to re-assert some control on output and prices.

What were the three main ways in which the major oil consumers responded to these price increases?

1   Production from non-OPEC sources of oil has been stepped up in various ways, e.g. by developing improved recovery methods for crude oil and by developing new fields, such as those in Alaska and under the North Sea.

2   Conservation measures including speed restriction on roads (e.g. 50 m.p.h. in the USA) and the disincentive effect of highly priced and taxed fuel (particularly in western Europe) have reduced the growth of oil imports.

3   Alternative sources of energy have been evaluated with renewed vigour.

2.2   Global energy requirement: a numerical perspective

Now that energy resources are transported more widely and in greater volumes than ever before, the accent is on international co-operation and planning. We need realistic quantitative estimates of the energy demand for the foreseeable future against which the sources of energy supply may be balanced. But as you read on, bear in mind that the field of energy resources changes faster and has more impact on economies than that of any other physical resource.

Historically, there has been an exponential increase in world demand for energy. This is partly reflected in the data used to compile the graph of energy supply and demand from 1950 to 2010 shown in Figure 4.

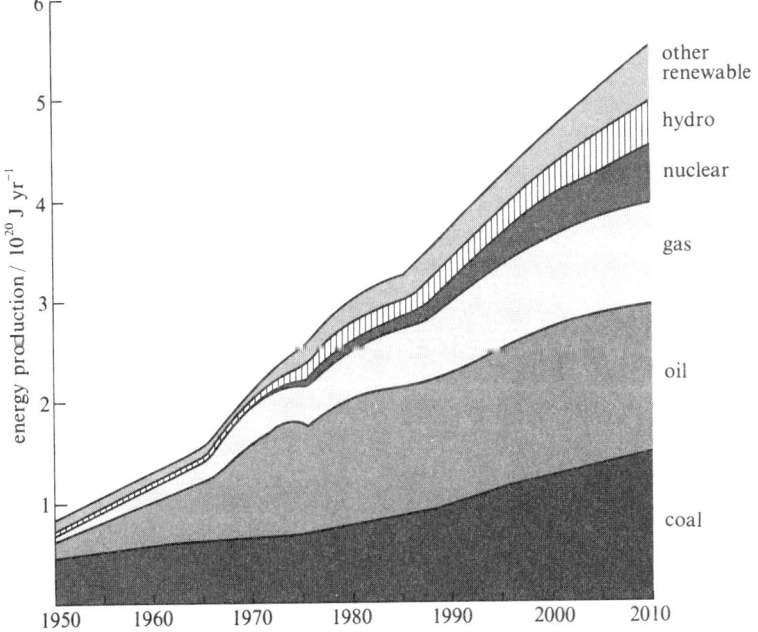

Figure 4   World energy production from 1950 to 1980, with projections for the future use of each resource category from 1980 to 2010.

What was the *doubling time* in the growth of total energy demand before about 1975?

Demand was running at about 10^{20} J yr^{-1} in 1954, increasing to 2×10^{20} J yr^{-1} by 1969; so the doubling period was approximately 15 years, which represents an annual growth rate of around 4.5 per cent.

If demand had continued at this level to the end of the century there would have been two doubling periods between 1969 and 1999 and so continued exponential growth would require a production figure of 8×10^{20} J yr^{-1} for 1999. The value shown in Figure 4 for 1999 is just under 5×10^{20} J yr^{-1} which implies a fall in the growth rate during this period.

It is clear that the days of abundant cheap oil are over. OPEC was established to try and control the price of petroleum by restricting its supply since for most comodities it is global economic forces that determine the fate of physical resources. In the long term historical sense, and considering physical resources as a whole, that is a fair enough generalization. The role of oil in recent years, however, shows how some physical resources can have a direct and far-reaching effect on global economics, and provides a notable exception to the generalization. From the 1950s to the early 1970s, the world became progressively dependent on cheap oil: a price rise was inevitable, but the rapid *rate* of the rise was such that it had a profound and global economic impact. The estimates of *projected energy demand* for 1985 shown in Figure 4 were made in 1980 and therefore and fairly certain but, of course, those for 1985 to 2010 are only the best informed guesses that were available at the time of writing in 1985. Nevertheless, as the above demonstrates, estimates made in 1970 of energy demand for the early years of the next century would have been almost twice that shown in Figure 4.

What happened to change these forecasts?

Of course, the increased cost of energy resources, particularly oil, during the 1970s is at the heart of the matter. If you look carefully at Figure 4 you will see that the upwards curve of exponential growth stopped in about 1975 when there was a sharp dip in oil and gas production. Apart from a period of low growth during the recession of the early 1980s the remainder of the graph has a *linear* rather than an exponential path, reaching 5.5×10^{20} J yr^{-1} by the year 2010. Despite the reduced growth rate, however, demand for energy has continued to rise and is expected to do so in the foreseeable future. This raises the question of how that demand will be met and the subsidiary components of Figure 4 indicate one way in which this might happen. Oil and gas are predicted to stay at their early 1980s level of production until 2010, whereas production from coal, nuclear power and hydroelectric power and other 'renewables' (winds, waves, solar power, etc.) is predicted to increase.

Can the fossil fuel bank sustain this demand? Given that the production before 1950 amounted to 25×10^{20} J and we can estimate production from 1950 to 2010 as 150×10^{20} J, we get a total of 175×10^{20} J. The fossil fuel bank is estimated to contain 10^{23} J (Chapter 1.1.2) so 17.5 per cent of this total will have been used by 2010.

Such an answer suggests that there are ample fossil to maintain the production levels indicated in Figure 4 until the year 2010; indeed a production level of $4 \times$

10^{20} J yr^{-1} from these sources could, in theory, be maintained for a further 200 years. But, as you will find in Chapters 3–9, matters are not that simple because (i) the vast bulk of the fossil buel bank is locked up in coal reserves, the production of which is difficult to accelerate, whereas oil and gas are predicted to produce more energy than coal up to 2010; (ii) significant improvements in oil and gas extraction technologies are required to meet even the targets up to 2010 set by Figure 4; (iii) the projection for each type of fossil fuel is related to the problem of what *form* future energy requirements will take, and the latter will be dependent on political decisions.

The last point raises several important questions. For example, what proportion of the fossil fuels will go to produce electricity, in which about 70 per cent of the energy production potential is lost in power stations that are only 30 per cent efficient? what proportion will go to driving cars, trains, aircraft, ships and so on? If oil is to be used as a fuel, for how long can it continue to supply the petro-chemicals industry? Can we actually locate and extract the necessary reserves in the vast amounts and at the rates suggested by the projections in Figure 4? Finally, to what extent can we reduce our dependence on fossil fuels? Can other sources of power expand tofill the *'energy gap'* between total demand and fossil fuel availability? These are just some of the questions you should bear in mind as you study this book. Now that we have traced the ancestry of energy resources and considered some projections of energy requirements, we can conclude Chapter 2 with a brief introduction to the geology of energy resources.

2.3   Energy resource distribution: a geological perspective

In common with most other highly prized physical resources with low *place value* resources of energy capital are unevenly distributed around the world (Figures 12 & 49) and the reasons are primarily geological. Certain critical combinations of environmental and tectonic conditions were necessary for these resources to have formed.

For example, to take a theme that is developed in Part II, the minimum *grade* (concentration) of uranium ore deposits worked today for nuclear fuel (*ca.* 350 p.p.m.) represents an enrichment of more than 100 times compared with the average crustal abundance of uranium (3 p.p.m.) As you will find in Part II there are several important categories of uranium deposits, some formed by surface processes controlled by changes in the acidity and oxidation potential of the water in sedimentary formations whereas others, formed by magmatic or *hydothermal processes,* occur in crystalline rocks as mineral infillings in fissures, joints and faults. The range of categories means that economic uranium deposits occur in various strata of different ages, the main criterion being that the crust must have been through several phases of magmatic or tectonic activity during each of which uranium became more concentrated.

In contrast, fossil fuels are produced under more specific geological conditions, through the anaerobic decomposition of plant and animal matter. Favourable conditions for oil and gas formation occur in marine sedimentary environments where biological production and sedimentation rates are high. On the other hand, coal represents continental accumulations of vegetation, frequently in near-shore equatorial regions subject to flooding by the sea. For fossil fuel formation, the *time* factor may be as important as the geological conditions, first because abundant organic matter derived from land was not available until about 400 Ma ago and secondly because prolonged burial is needed to form certain categories of depostis. Figure 5 summarizes the distribution of energy capital resources with respect to geological time.

You should note several significant features in connection with Figure 5:–

1   The most productive geological periods for oil abundance are from 190 Ma to the present although small amounts may be found in older rocks. Oil is most likely therefore to be found in off-shore sedimentary basins of Mesozoic or Tertiary ages.

2   Coals are not found in rocks older than 400 Ma because there was no abundant plant life on land before that time. Coal deposits occur in low lying continental swamp areas frequently of Carboniferous or Permian age.

3   Natural gas is associated with the formation of both oil and coal and consequently it may be found in rocks yielding either of these fossil fuels.

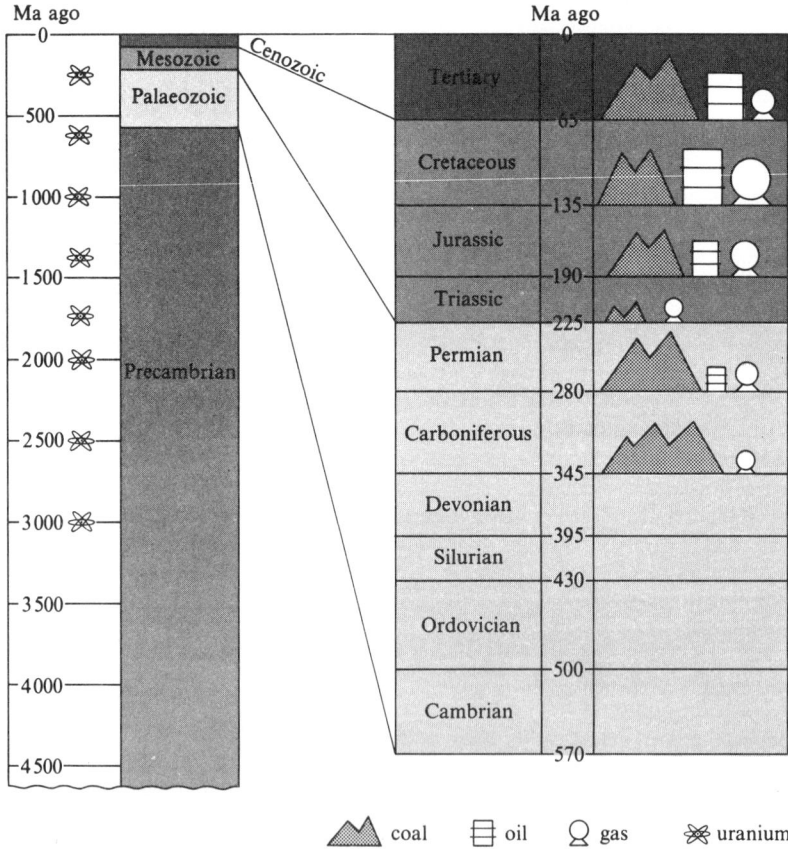

Figure 5   Schematic illustration of the relative abundance of different energy resources in different geological periods.

4   In contrast uranium may be found in either sedimentary or crystalline rocks of all geological ages which lie within ancient continental shield areas (or cratons).

5   Whereas almost one fifth of the fossil fuel bank is likely to have been used by 2010, the bank took 400 Ma to accumulate.

2.4   Summary of Chapter 2

1   Historically, wood, wind and moving water have been the most widely used energy resources. But since the Middle Ages, first coal, then oil and natural gas, have become progressively much more important sources of energy. At the time of writing over 80 per cent of global energy production is based on the use of fossil fuels and this will probably remain true at least until the turn of the century.

2   Following these growth patterns of coal and then oil (plus natural gas) production, nuclear power may become the modern growth industry, although much attention is also being devoted to renewable energy sources.

3   Since the late 1950s the supply and demand of energy resources have become of international concern with important political overtones; hence international organizations as the OPEC cartel have been established.

4   The geological processes that produce energy capital resources depend on critical combinations of environmental and physical conditions coupled with time constraints; for example, coal accumulations only began with the evolution of terrestrial plants. As a result, energy resources are markedly uneven in global distribution and this had been reflected particularly in the sensitivity of oil supply from Middle Eastern sources and the demand for oil in the western world.

5   Rapid increases in oil prices, during the 1970s produced a total re-evaluation of energy policy in the western world and, in particular, halted exponential growth in the world energy demand, though this demand continues to increase linearly and is projected to reach 5.5×10^{20} J yr^{-1} by the year 2010.

3 Geology of coal deposits

Study Comment. Chapter 3 deals with the geological events that led to the accumulation of plant material and resulted in the occurrence of coalfields in many regions of the world. An understanding of the characteristics of different types of sedimentary rock and the various environments in which they accumulated is needed to appreciate the significance of these geological events, in particular the progression of coal maturation and the chemical properties associated with each stage.

3.1 Origins of coal

Coal is formed from the fossilized remains of land plants, and its organic origin can be recognized from the presence of various plant fragments preserved in sediments associated with beds of coal, or *coal seams*, such as leaf fronds (Figure 6), branches and roots. Occasionally, fossilized tree stumps are found preserved in life positions.

Figure 6 Fossilized leaf *(Pecopteris)* in a Coal Measures deposit (roughly actual size).

The precursor of coal is *peat*, which is familiar as the compressed soggy plant remains that characterize the bogs and swamps of poorly drained regions. Where conditions are suitable, peat deposits can accumulate to considerable thicknesses and there are extensive areas of peat in the moorland districts of the British Isles. For centuries peat has been cut into blocks, dried and used as a domestic fuel, and in Eire it is used to fuel power stations; it is also extensively used in the horticultural industry.

The first prerequisite for the formation of a peat deposit is that large amounts of plant debris have accumulated and are preserved because the ground is saturated with water, i.e. the *water table* is virtually at the surface. For organic matter to be preserved, non-oxidixing or anaerobic conditions must apply to prevent the reversal of the photosynthesis reaction. Swamps and bogs provide such an environment and can form over a wide range of latitudes, from tropical rainforest to the high latitude tundra regions.

The second essential requirement is that the geological setting can accommodate great thickness of peat, because it takes about 10 metres of plant debris to yield 1 metre of coal. For such compression to occur, the peat has to be buried beneath a considerable thickness of sediments. This readily occurs in low-lying estuarine and coastal regions, provided that two conditions are fulfilled: first, there is overall subsidence of the region, and secondly, there are continuing supplies of sediments such as sands and muds to bury the peat. Flooding of the swamp will end peat formation with its burial beneath sediments, but a drop in the water table allows the peat to dry out and be removed either by erosion or oxidation resulting from forest fires. Distinctive layers of charcoal identified within coal deposits indicate the widespread extent of such oxidation.

3.2 Sedimentary environment of coal deposits

The ratio of the total thickness of coal seams in a sequence to that of the enclosing sedimentary strata is commonly very small, between 1:10 and 1:100. In contrast to the slow and stable deposition of the coal-forming peat, the sedimentation of the great volume of associated strata was relatively rapid and variable, representing deposition under various geological conditions. You can understand these variations when you examine Figure 7, which shows a typical sequence of sediments found in the British coalfields, and indeed in coalfields in many parts of the world. The following features are important:

1   The coal seam is often underlain by a *seatearth* , so-called because it is believed to represent a fossil soil in which the plants grew, i.e. the 'floor' of the peat bog. Carbonaceous markings, believed to be the rootlets of the plants, are one characteristic, and the chemical composition indicates that seatearths were leached of soluble elements (potassium, sodium and calcium, for instance) by the acidic waters of the swamps. They can be clays or sands and are often a source of useful raw materials for the construction industry. Some are *fireclays* or refractory clays, which are low in *fluxes* and therefore have high melting points. The quartz-rich sands are known as *ganister* and are sources of both industrial sands and refractory bricks used in industrial furnaces. The shales and clays associated with coals are a widely used source of brick clays in Britain.

2   The coal seams are generally overlain by muddy sediments which pass upwards into sands and then to seatearths and another coal seam. Coalfields may be characterized by sequences of seatearths, coal, shale and sandstone. These cyclic sequences are repeated time and time again, although, in any sequence, parts of a cycle may be missing or duplicated, while the thicknesses of the different layers can vary greatly.

3   Occasionally there may be a shale with distinctive marine fossils known as a *marine band*, these normally occur above the coal seam (Figure 7). For the most part, however, the fossil remains associated with coal-bearing sequences are of freshwater or terrestrial species and so we know that the original peat swamps accumulated in freshwater conditions.

While rootlet beds indicate fossil soils, other plant remains are less diagnostic as they may have been transported by water away from their place of growth. Coal-bearing cycles are well known in the Carboniferous coalfields of Europe and North America, and they are widely recognized in other formations. Studies of individual cycles in Britain have shown that they vary considerably across a coalfield. While swamp conditions with little sediment input may have existed at one locality, sands, silts or mud were being deposited elsewhere. All these sediments were formed in shallow water conditions, yet the sediment thicknesses in coalfields reach hundreds or thousands of metres.

Consideration of all these features and the application of the *Principle of Uniformitarianism* have led geologists to conclude that the environments in which the coal swamps developed were similar to present-day coastal swamp or *deltaic environments*, especially those forming in areas such as the Mississippi and Niger deltas. Deltas develop at the mouths of rivers where sea currents are unable to remove the sediment transported by the river. Sediments are deposited in a typically fan-shaped lobe as the river divides into a number of smaller channels, known as *distributaries*, which meander across the flat surface of the delta plain.

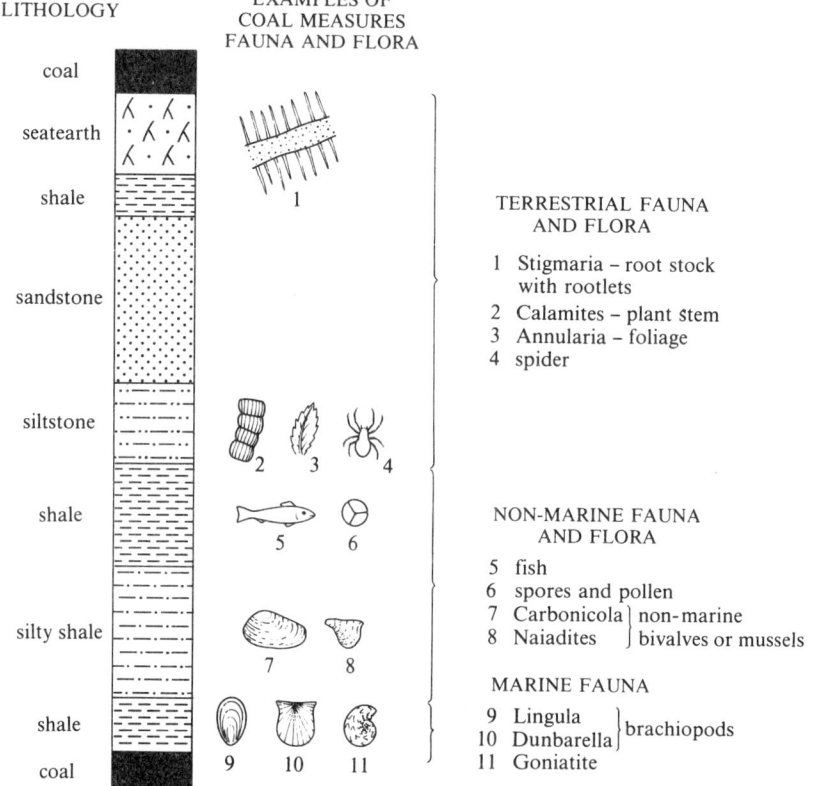

Figure 7 An example of a cyclic sequence in a Coal Measures deposit, showing the variety of sediments and a selection of typical fauna and flora.

The delta plain builds up by successive accumulations of muds, with silts deposited at times of flood. It is characteristically swampy and covered with vegetation, perhaps leading to the accumulation of peat. The distributary channels become progressively filled with sandy deposits, which are preserved as irregular 'ribbons' of coarse sediment, more or less lens-shaped in cross-section, amongst the predominant muds and silts (Figure 8). There is thus considerable lateral variability in the sediments of a delta. Only the marine bands are continuous throughout the delta as they were deposited by seas which inundated the whole area. They therefore provide important *marker horizons* in otherwise highly variable sequences.

Boreholes sunk into the delta plain reveal cyclic sequences, like the one in Figure 7. This is because deltas subside intermittently and irregularly, with each phase of subsidence initiating a new cycle of deposition that culminates in the development of potentially peat-forming swamps.

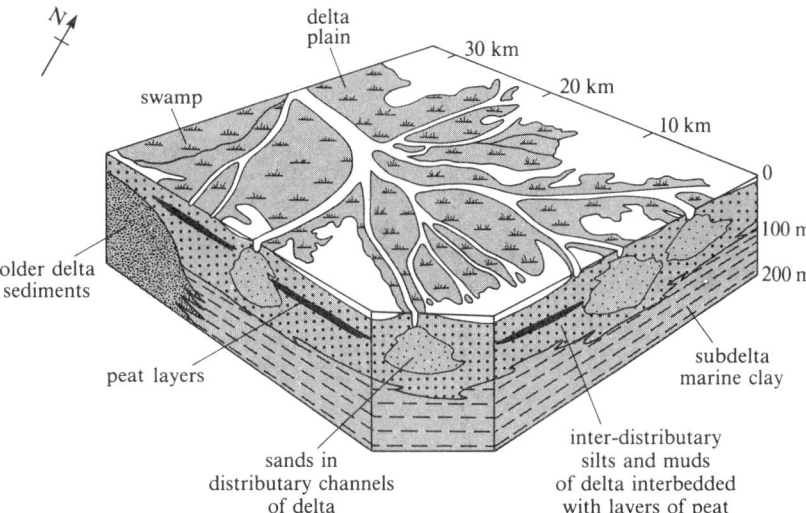

Figure 8 Features of the Mississippi delta. The diagram shows elongated lenticular sand bodies formed by distributaries in the inter-distributary sediment of carbonaceous silts and muds. The vertical scale is exaggerated 30 times.

Coal seams can vary in thickness from mere films to several metres. Most of the seams worked in British coalfields are between 1 and 2 metres thick, but in other parts of the world much thicker seams are worked.

Nearly every seam in the British coalfields either splits into two or more beds separated by bands of carbonaceous shale known as *dirt bands*, or converges to join with an adjacent seam (Figure 9).

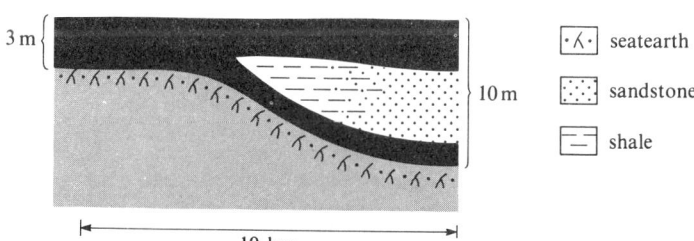

Figure 9 An example of splitting in a coal seam.

3.3 Coal formation

The first stage in coal formation is one of *biochemical decomposition*. As dead plant material accumulates in the swamps, anaerobic micro-ogranisms such as bacteria and fungi start to break it down. Plants are largely composed of cellulose (a carbohydrate forming cell walls) and lignin (the woody material of plants). The process of decomposition involves the breakdown of the more soluble components, principally the cellulose, with a resultant enrichment of the lignin and the waxy constituents from leaf coatings, spores, pollen, fruit and algal remains. The decomposition results in the loss of volatile constituents, such as carbon dioxide, water and methane, leaving compacted structureless compounds enriched in carbon.

The second phase of coal formation is *coalification* or *maturation*, which commences when the accretion of the peat deposits is arrested by burial beneath mud, sand and silt. Coalification involves physical and chemical changes imposed by increasing temperature, pressure and time. The stages are known as levels of *rank*, a term that is used to indicate the maturity of the coal. The different rank stages, *peat, lignite, sub-bituminous, bituminous, anthracite* and *graphite*, together with some of the parameters used to define them, are listed in Table 2. Changes in rank are gradual and so the boundaries of the rank categories are arbitrary ones, denoted by the dashed lines in the first column of Table 2.

The continuous variation of chemical composition through the rank series is shown in Figure 10. The most important chemical change is the increase in carbon at the expense of oxygen. The proportion of hydrogen present remains relatively constant at 4–10 per cent by weight over much of the rank series until about 90 per cent carbon, where a significant reduction in the amount of hydrogen commences. As the rank is enhanced the porosity of the coal decreases and as a result its density increases. The appearance of the coal also changes from dull when low rank to very shiny and metallic when of high rank.

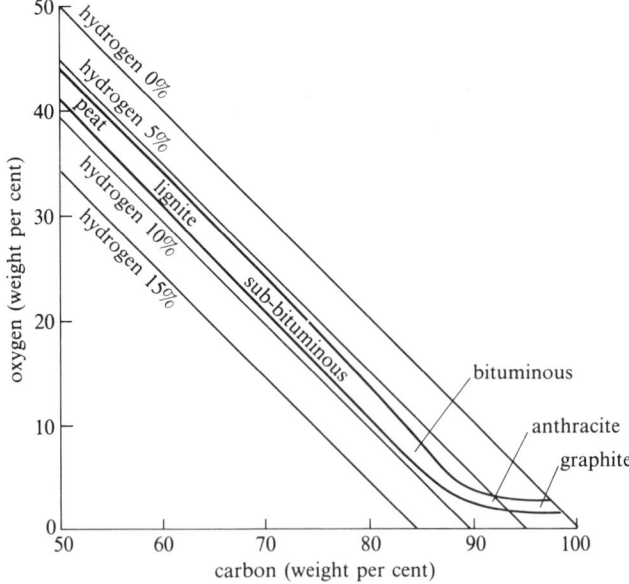

Figure 10 A graph showing the relationship between carbon, oxygen and hydrogen contents for the six stages in the rank series.

Table 2 Changes in carbonaceous material and its chemical composition at different rank stages

Rank	Description	Typical content of volatile matter*, ** (weight per cent)	Typical elemental analysis* (weight per cent)		
			Carbon	Hydrogen	Oxygen
peat	decomposed fibrous material	more than 50	less than 50	7	more than 35
soft brown coal or lignite	still contains woody material		60	6	25
	hard blocky appearance	45			
hard brown coal or sub-bituminous coal	dark material developing slight lustre		75	5	15
bituminous coal	black, lustrous and strongly banded appearance	35	85	5	10
anthracite	metallic lustre; bands absent	10	90	4	5
graphite	very hard, with a silvery lustre	less than 5	more than 95	less than 4	—

*These values are calculated on the basis of pure coal substance (i.e. free of moisture and ash).

**Volatile matter* is a mixture of combustible gases (mainly hydrogen, carbon dioxide and methane) and condensible substances given off when coal is heated in the absence of air.

The changes summarized graphically in Figure 10 are due to the expulsion of water vapour, carbon dioxide and methane. Only small amounts of methane (CH_4) are liberated during the early stages of coalification, but during the transition from bituminous coal to anthracite (particularly over the range 85–92 per cent carbon) much hydrogen is eliminated as methane and other hydrocarbons, whereas the emission of carbon dioxide declines. Much of the methane has migrated into overlying rocks, and some of the world's *natural gas* fields, such as those in the southern North Sea (see Chapter 9.2) are accumulations of methane derived from the vast coalfields that underlie them. The expelled methane is the explosive gas 'firedamp', which can be so dangerous in mines that produce coal of high rank. When coal is extracted methane is released into the mine air and adjacent strata with the consequent risk of explosions.

Another way of classifying coals is according to the type of organic matter from which they were formed. Under the microscope coal can be seen to consist of individual components called *marcerals*, which include *vitrinite*, the remains of woody tissue; *inertinite*, the charcoal-like remains resulting from bacterial oxidation of plant material or from ancient forest fires; and *exinite*, the waxy remains of spores, pollen, leaf coatings and algae. The great majority of coals are *humic coals*, formed from woody tissues and so the maceral types vitrinite and inertinite are predominant, forming the bright vitreous bands in bituminous coal. Much less common are the *sapropelic coals*, which are dull and unlaminated in appearance. They were formed from fine-grained material, including the waxy residues from spores and algal remains, that accumulated as organic muds in sheltered lakes and lagoons and did not pass through a peat stage; exinite is the dominant maceral type in this class of coal. They can be sub-divided into *cannel coals*, which are rich in spores, and *boghead coals*, rich in algal remains. The maturation paths of humic and sapropelic coals are summarised in Figure 11.

3.3.1 Causes of rank change in coal

Geologists have long recognized that rank increases with depth in an extended sequence of coals. Temperature increases with depth of burial because of the *geothermal gradient*. Pressure is obviously important in compressing the peat and this of course increases with depth of burial and with *orogenesis*. Time is also important: in general, the longer a coal is buried, the higher its rank is likely to be. Temperatures of 150–200 °C are sufficient to form high rank coal when they persist over very long periods and the depth of burial was sufficient. However, if the depth of burial was too shallow and the temperature did not rise then the peat will not be converted into high rank coal, no matter what its age. Thus, for example, Carboniferous coals near Moscow, 340 Ma old, are still brown coals (lignite) whereas Cretaceous coals in the Canadian Rockies, 120 Ma old, are anthracitic.

3.4 Properties of coal

The outstanding geochemical characteristic of coals is the remarkably high content of carbon and, to a lesser extent, hydrogen compared with the proportions of these elements in other types of sedimentary rock; this is a reflection of the biogenic origin of coal. The relative proportions of the major constituents of coals change with increasing rank (Table 2 and Figure 10), and these changes affect their properties and their suitability for different uses.

1   Coals rich in volatile matter (more than 30 per cent) are easy to ignite and burn freely but with a smoky flame. Low volatile coals are more difficult to ignite but they burn with a smoke-free flame, providing a natural smokeless fuel.

2   Coking properties: the carbonized residue remaining after the volatile matter has been driven off in the absence of air is called *coke*. Coals in the range of 85–89% carbon content (high rank bituminous) become partly fluid on heating and swell up to form a porous coke. These *coking coals* are especially valuable for the coke used by the iron and steel industries. Being free of volatiles coke also provides a useful (artifical) smokeles fuel.

3   The *calorific value* of coal is the amount of heat liberated under standard conditions and is usually expressed in Kilojoules per gram (kJ g^{-1}). Calorific value generally increses with rank, ranging from 15–26kJ g^{-1} in low-rank lignites through 31–35kJ g^{-1} for bituminous coals to 30–33kJ g^{-1} in anthracites.

4   *Reflectance* : a convenient method of determining the rank of coal is to examine highly polished surfaces with a microscope using reflected light. Amongst humic coals there is a general increase in reflectance with increasing rank — higher rank coals have a greater 'shine' than low rank coals — and the increase in reflectance of vitrinites can be accurately measured and related to rank.

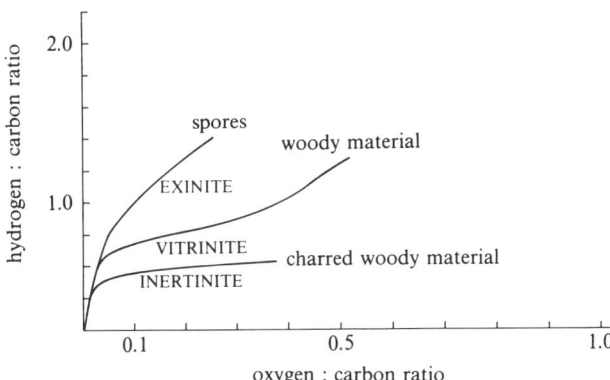

Figure 11   Maturation diagram for coal showing the composition of various types of terrestrial organic matter and of the three main coal macerals. The maturation path of humic coal in represented by the pathway of woody material and charred woody material to vitrinite and inertinite; the maturation path for sapropelic coals is represented by the pathway from spores to exinite; coals of the highest rank, anthracite, would plot roughly where the three curves meet.

3.4.1   Impurities in coal

The inorganic impurities within coal seams contribute to the residue or *ash* left after combustion. This mineral matter either accumulated with the original plant material or was precipitated during coalification. Its distribution within the seam is important for practical reasons, determining the extent to which the coal can be cleaned before sale. The melting temperature (fusibility) of the ash and its tendency to corrode boiler surfaces are also dependent on the composition of the mineral impurities. These are important considerations when assessing the suitability of coals as boiler fuels or for coke making.

What minerals would make up the *clastic sediments* deposited with the original plant debris?

They would be fine muds washed into the accumulating peat deposits and consist of clay minerals and fine quartz particles. When the coal swamps were formed, Britain lay in tropical latitudes and so we might expect the principal clay mineral to be *kaolinite*, and kaolinite does indeed occur. *Diagenetic* changes associated with coal formation, however, have converted much of this into *illite* which, together with 'mixed-layer' clay minerals, is the commonest type found with coal.

Next to the clays, the most important group of impurities is the carbonates. During biochemical decomposition and the early stages of coalification, carbonate minerals are precipitated by diagenetic processes either as *concretions* (hard, generally disc-shaped nodules up to tens of centimetres in size) or as infillings or veins within fissures in the coal seams. They are composed of the minerals siderite, $FeCO_3$, and ankerite, $Ca(Mg, Fe)(CO_3)_2$, and the fusibility of the ash is strongly influenced by the total carbonate content and the proportions of calcium, magnesium and iron. Siderite concretions can also form extensive bands in shales and they provided an important source of iron ore in Britain during the early stages of the Industrial Revolution.

Sulphur is another common impurity found in all coals, in proportions varying from 0.5 to several per cent, and is principally present as pyrite (FeS). On combustion it yields sulphur dioxide (SO_2), which causes corrosion in boilers and contributes to atmospheric pollution.

Another important impurity is sodium chloride (NaCl). When it is present in high concentration (over 1 per cent chlorine content) the coal is virtually unusable in power stations because of severe boiler corrosion. Coal also contains minute amounts of a large number of *trace elements*, as do other sedimentary rocks; a few elements may show enrichment in coal, including germanium, arsenic and uranium. Although most of these end up in the ash after combustion, it has been alleged that some trace elements released by burning coal may contribute to atmospheric pollution.

3.5 Ages of coal deposits

You already know from Section 2.3 that there were no coals laid down before about 400 Ma; the first important coal deposits occurred in the Lower Carboniferous 345 Ma ago. The earliest simple land plants appeared first about 400 Ma ago, before which the land masses lacked any significant vegetation cover; it was not until the explosive evolution of land plants at the beginning of the Carboniferous that the accumulation of plant remains to form coal deposits became posible.

Two major periods of coal formation can be recognized (cf. Figure 5), first in the Carboniferous to early Triassic, 345 to 200 Ma ago and then during the late Jurassic to early Tertiary, 150 to 50 Ma ago. The younger coals are mainly brown coals and lignites whereas the older coals are mainly bituminous coals and anthracites, although there are exceptions to this generalization (we cited two in Section 3.3).

Figure 12 shows the worldwide distribution of coal deposits, and it is clear that

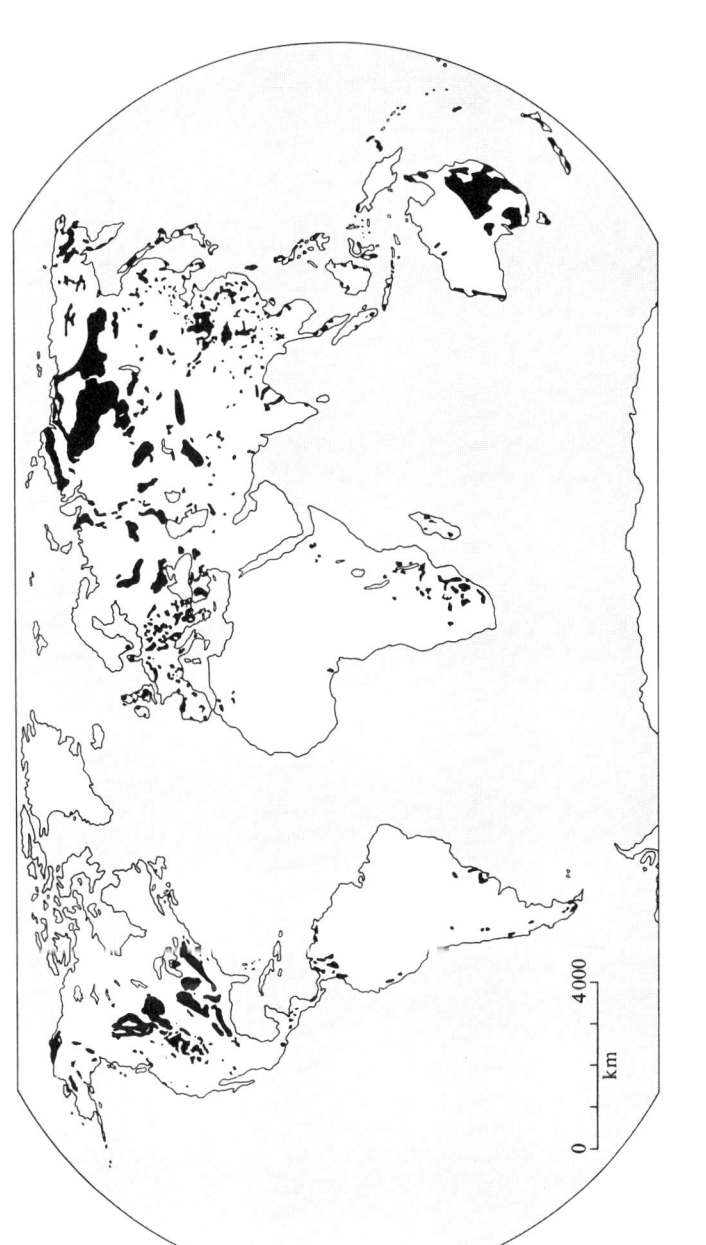

Figure 12 A world map showing the distribution of known coal deposits.

the known major areas are principally in the northern hemisphere. The nations of the Middle East and the southern continents appear to be deficient in coal deposits, with the exception of Australia. However, for Africa and South America this may reflect to some extent the lack of exploration for coal. There are two broad belts of coalfields from the older, Carboniferous – Triassic Periods: a broad chain of large coalfields of Carboniferous age extends from the USA through western and eastern Europe, the USSR and into China; a second chain of Permo-Triassic coalfields is found in the southern continents — South America, southern Africa, Australia and Antarctica.

The fossilized plant remains found with the Carboniferous coals suggest that they were formed in tropical swamps. For example, the virtual absence of growth rings in fossil logs is indicative of minimal seasonal changes, such as occur in equatorial zones.

Why then are most Carboniferous coals now found in temperate latitudes?

The Earth's climatic belts are not thought to have changed through time but there is much evidence to suggest that the continents have moved. *Palaeomagnetic* determinations support the theory of continental drift and show that during Carboniferous and Permian times the northern continents lay in or near equatorial latitudes.

Figure 13 shows a reconstruction of a continental configuration at the close of the Permian period, some 225m years ago, when ocean closure and continental collision led to the unification of all the main masses of continental crust into one super-continent termed Pangaea. Britain was then located some 10° North of the equator.

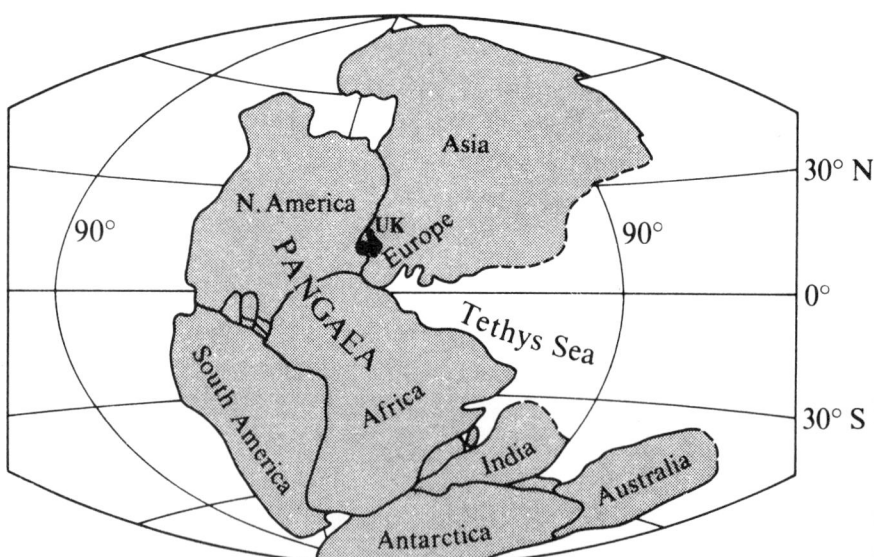

Figure 13 Reconstruction of the continental configuration at the end of the Permian period 225 ma.

The Permo-Triassic coals of the southern continents are rather different in appearance with a different flora and clearly defined growth rings, consistent with formation in temperate latitudes. The relative abundance of charcoal-like material is also indicative of drier and temperate conditions.

Britain was still near equatorial latitudes in Permian times, as was much of Europe, and the major *evaporite* deposits of the northern hemisphere were deposited at the time. Figure 5 indicates that important coalfields also occurred in many younger rocks — these are the Mesozoic — Tertiary lignites and brown coals found principally in the USA, Canada, Western Europe, the USSR, China and Australia and we can deduce that most of these coal deposits were formed in tropical or sub-tropical coastal swamp conditions.

3.6 Summary of Chapter 3

1   Coal formed from the fossilized remains of land plants, preserved after death in the anaerobic environments of swamps. These swamps can occur at many latitudes and represent large areas of poorly drained, low-lying land.

2   A typical deltaic coalfield consists of cyclic sequences of sediments. There may be hundreds of cycles, and these show marked lateral variations. One important economic consequence of these variations is that individual coal seams are likely to vary both in thickness and in lateral extent.

3   The transformation of peat to coal involves two distinct phases — a first stage of biochemical decomposition, involving degradation by anaerobic micro-organisms, and a second extended phase of coalification imposed by increasing temperatures and pressure continued over tens to hundreds of millions of years. Six stages of coalification are recognized in the rank series, each characterized by particular chemical and physical properties. Several deleterious inorganic constituents also occur in coal; for example, high contents of sulphur or chlorine which cause corrosion in boilers.

4   The oldest important coalfields are of Carboniferous age. Their deposition followed the rapid evolution and spread of land plants some 350 Ma ago.

5   The coalfields of the northern hemisphere are generally Carboniferous and were formed in tropical latitudes. Those of the southern hemisphere are largely Permian or Triassic and originated in temperate latitudes.

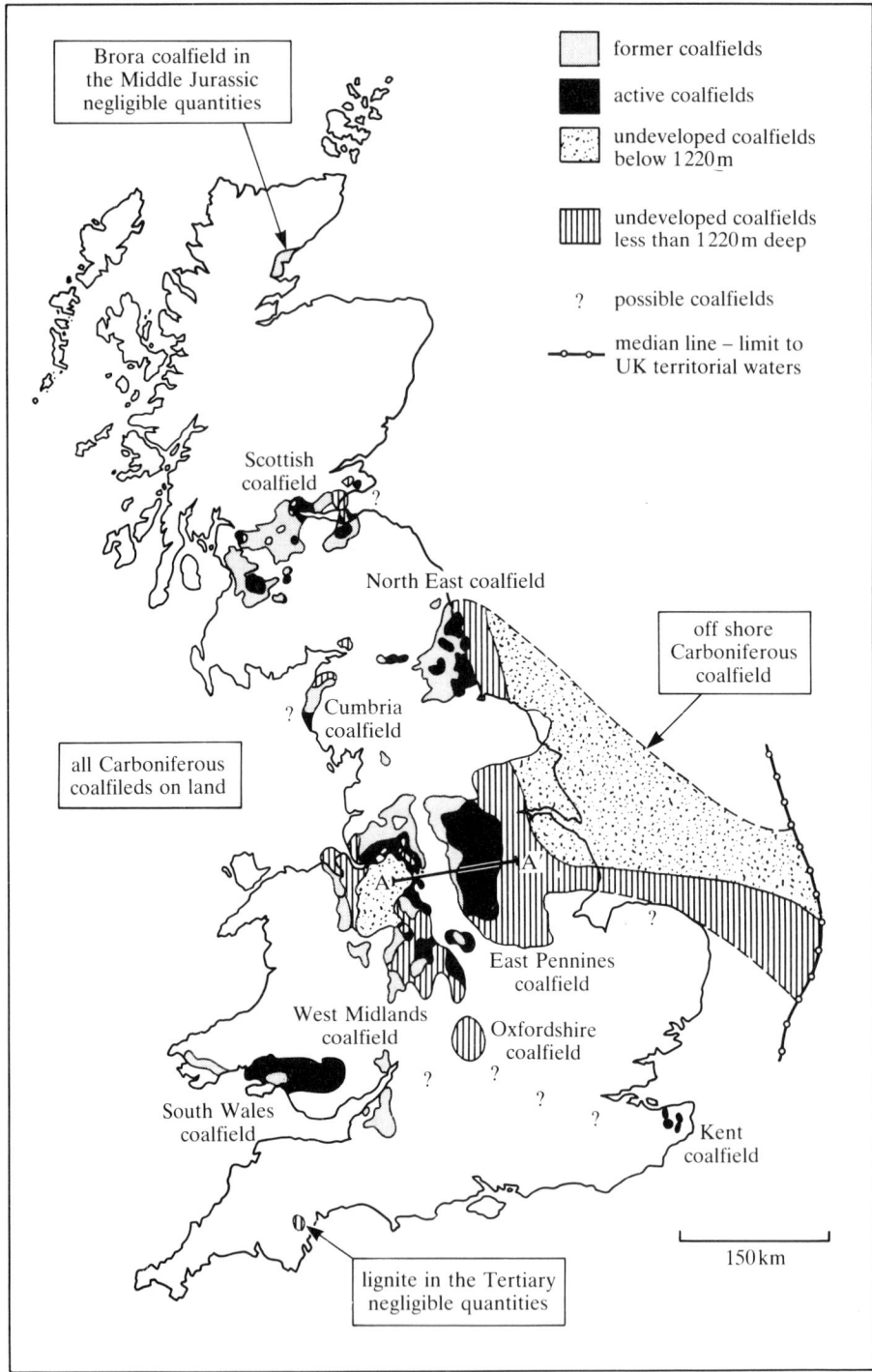

Figure 14 The location of the British coalfields and their offshore extension in the southern North Sea.

4 Coal: exploration and extraction

Study Comment. In this Chapter we examine the geological, technical and economic constraints that control the coal mining industry and are likely to determine its future. First we look at the locations of the principal British coalfields. Coal mining is then discussed under three headings — exploration techniques, surface mining and underground mining. Finally, we consider how the coal is used and how attempts have been made to evaluate resources for the future.

4.1 Locations of the coalfields

Although coal mining in Britain commenced in the coastal coalfields of Scotland, north-east England and south Wales, the resources of these regions are now becoming exhausted and mining activities there are declining. Figure 14 indicates that the principal working coalfields are now located in the English Midlands, particularly in Yorkshire, Nottinghamshire, and Derbyshire, an area known as the East Pennines Coalfield. Compare Figure 14 with a geological map of Britain to see the setting of the coalfields.

All coal mined in Britain today comes from rocks of Carboniferous age. Younger coals do occur in Britain, but they are not of any economic importance. A small Jurassic coalfield, part of a larger off-shore field, was formerly worked at Brora on the north-east coast of Scotland, and small deposits of Tertiary lignites have long been known at Bovey Tracey in south Devon. In 1982 important Tertiary lignite deposits were discovered on the eastern shores of Lough Neagh in Northern Ireland which may become significant to the local economy.

Let us now consider how the Carboniferous coalfields were formed. Figure 15 shows the approximate distribution of land and sea during the Upper

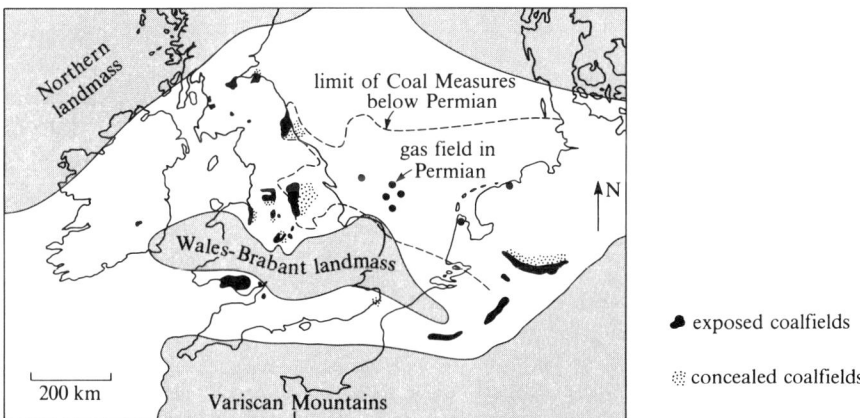

Figure 15 The distribution of the coalfields of north-west Europe, together with the area of the southern North Sea underlain by Coal Measures. This vast area was originally covered by Carboniferous deltas which originated from the Northern landmass. The Wales—Brabant landmass persisted as a land area throughout most of the Upper Carboniferous, separating the northern and southern deltaic provinces.

Carboniferous, along with the major coalfields. The boundaries show average positions of shorelines at the time, although there were considerable fluctuations in their positions owing to advances and retreats of the sea.

Comparison with a geological map will show you that the ancient rocks of north-west Scotland and north-west Ireland could have been areas of land throughout the Carboniferous; another area of land centred on the Lower Palaeozoic rocks of Wales, extending westwards to Ireland and eastward to include the pre-Carboniferous rocks beneath the sediments of eastern England, is called the Wales—Brabant landmass (Figure 15). During the Lower Carboniferous, most of the white areas of Figure 15 were covered by shallow seas in which Carboniferous limestone deposits were laid down; the exception was in what is now south-west England, where shales and sandstones were deposited in deeper water. Then a widespread delta complex extended south from the Northern landmass into these shallow seas, and coarse-grained sandstones, such as the Millstone Grit, were deposited.

Later in the Carboniferous, coastal swamp conditions developed over vast areas and were covered by extensive forests. Great thicknesses of anaerobically rotting vegetation accumulated in layers separated by thick deposits of muds and sands in a deltaic environment very similar to that illustrated in Figure 8. Other deltas developed along the southern flanks of the Wales–Brabant landmass resulting in coalfields today between south Wales and Kent.

However, the coalfields today are much less widely distributed than the former coal swamps. This is because at the end of the Carboniferous the area south of the Wales–Brabant landmass was subjected to intense north–south compression during the *Variscan orogeny*. As a result the coalfields south of the Wales—Brabant landmass, particularly in south Wales and Somerset, were folded into a series of east—west trending *anticlines* and *synclines* in which steep dips of 40° or more are not uncommon and which were subjected to considerable faulting. These are the principal reasons why the south Wales coalfield contains high rank coking coals and anthracites and at the same time is most difficult to mine. The Kent coalfield, which is linked under the English Channel with the Pas de Calais coalfield in France, is much less disturbed.

The southern basin, south of the Wales—Brabant landmass (Figure 15), provides the most intriguing possibilities for new coalfields within the British Isles. Proving of the concealed Kent coalfield (Figure 15) beneath Mesozoic and Tertiary rocks in 1890 led in succeeding years to speculation on the possibility of other concealed coalfields in southern England. In 1960–61 the Institute of Geological Sciences sank an exploratory borehole near Witney to the west of Oxford and proved eight coal seams. Subsequent exploration has indicated that a substantial coalfield exists in Oxfordshire. We will consider the implications of this and other discoveries when we review Britain's coal resources.

North of the Wales–Brabant landmass the disturbances were much less intense and the Carboniferous sediments were folded into a series of broad basins separated by anticlines with gentle dips, rarely exceeding 10°. Subsequent erosion on a vast scale removed several hundred metres of Carboniferous strata from the anticlinal areas so that the central and northern coalfields are separated into isolated synclinal basins. This period of erosion at the end of the Carboniferous was followed by the deposition of the sandstones and limestones of the Permian and subsequent periods, concealing great areas of coal-bearing sediments beneath younger rocks. The *unconformity* between the folded and eroded Carboniferous rocks and the overlying younger rocks represents a vast time interval.

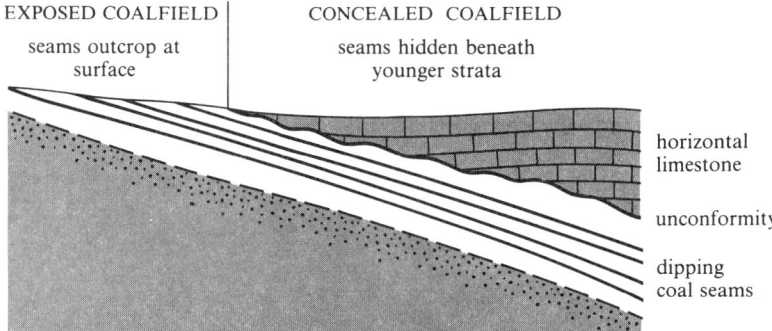

EXPOSED COALFIELD | CONCEALED COALFIELD

seams outcrop at surface | seams hidden beneath younger strata

horizontal limestone

unconformity

dipping coal seams

Figure 16 A diagram illustrating the distinction between exposed and concealed coalfields.

Coalfields can be divided into two categories — *exposed coalfields* where the coal-bearing rocks outcrop at the surface, and the *concealed coalfields* where they are hidden beneath unconformable younger strata (Figure 16). The exposed coalfields have been defined with considerable precision by surface mapping and earlier workings but the geological features of the concealed coalfield have to be located by a variety of exploration methods including boreholes and geophysics. Mining commenced in most coalfields in the exposed areas and then gradually extended into the deeper parts of the concealed coalfields as the shallower seams were exhausted. A good example is the East Pennines Coalfield, which produces the bulk of Britain's coal (Figure 14). The working commenced in the exposed coalfields of Derbyshire and west Yorkshire but are now largely concentrated in the deeper eastern regions beneath a cover of Permian and younger strata.

Figure 17 shows a cross-section from Cheshire eastwards to Lincolnshire, represented by line A—A' in Figure 14. You can recognize the deep synclinal basin of the Cheshire coalfield separated from the more extensive Nottinghamshire coalfield in the east by the Lower Carboniferous limestones of Derbyshire. The coal-bearing rocks in the Cheshire basin are buried very deep (1500–2500 metres) so large resources are inaccessible to current mining techniques.

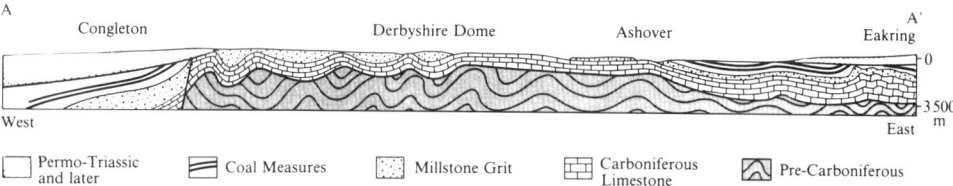

A

Congleton   Derbyshire Dome   Ashover   A'
Eakring

West   East

Permo-Triassic and later   Coal Measures   Millstone Grit   Carboniferous Limestone   Pre-Carboniferous

Figure 17 A geological cross-section (about 90 km) across the Pennines from the Cheshire basin to Lincolnshire, showing the relationship between the Carboniferous and overlying younger strata.

The coastal coalfields of north-east England and the Midland Valley of Scotland extend under the North Sea, where vast reserves have been proved from offshore drilling rigs. Undersea workings from some Durham collieries already extend over 9 km from the coast. Undersea workings have also been undertaken from collieries in west Cumbria and north Wales, but to date, the large offshore reserves adjacent to the Yorkshire–Lincolnshire coast have not been mined.

4.2 Exploration and evaluation techniques

Exploration for new coal reserves is an integral part of mining activity, indeed mining is itself an exploration process as well as a producing process. The need for exploration is covered by several factors:

1. the requirements of existing mining operations to maintain or increase production
2. the need to make up deficiencies in production due to unexpected and rapid geological changes
3. the proving of new reserves which allow the replacement of exhausting mines

There needs to be a balance between exploration for new mine sites and exploration to safeguard or increase production at existing mines. The greatest loss to the UK coal mining industry could be to miss out on the opportunities that exist to produce coal from existing mines fortunate enough to be sited where large potential reserves exist and where mining can be productive in the search for short-term profitability. During the 1980s therefore exploration has been principally concentrated on proving reserves for existing collieries, leaving further exploration of many potentially new mine sites for the future.

The two exploration techniques most widely used for coal are core drilling and logging and seismic reflection profiling. These techniques complement each other, with seismic profiling providing a relatively cheap means of surveying the ground between very expensive and widely spaced boreholes.

4.2.1 Drilling

The most effective exploration technique is drilling boreholes where a solid core of rock can be obtained. Recent coal discoveries were made possible by the improved drilling and core recovery techniques developed since the early 1960s. The secret of good core recovery is the modern *core barrel*. Figure 18 compares the simple core barrel used in the 1930s with the modern version. In the old core barrel (Figure 18a) the core revolved inside the drilling tubes as drilling proceeded. This frequently resulted in the friable coal cores breaking into fragments, which fell back into the hole when the drill string was removed and

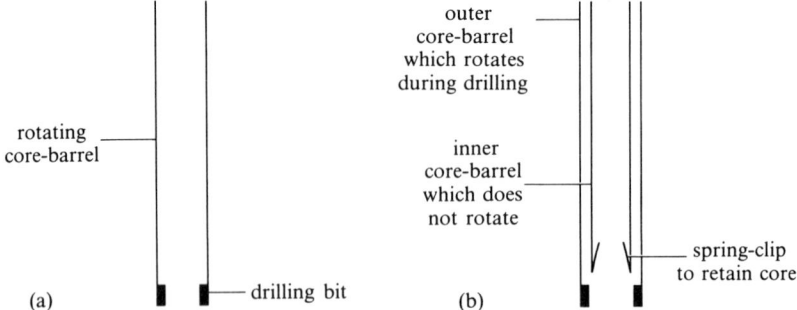

Figure 18 Diagrams illustrating developments in the design of core barrels used for coal drilling: (a) old core barrel; (b) modern core barrel. Barrel diameters vary depending on the drilling depth: typically they are about 10 cm.

were then pulverized as drilling recommenced. The modern drill tube contains an inner core barrel (Figure 18b), which holds the coal core by means of a spring clip but remains still as the drill revolves. A further refinement enables the core barrel to be raised inside the drill rods to the surface when full without all the rods having to be extracted — a procedure known as 'wireline drilling'.

Coring long sequences of strata is a slow and very expensive procedure and important segments of core can be lost in drilling. Cores have to be carefully examined by a geologist and the results recorded — a procedure known as *core logging*. Geophysical techniques have been developed to record the nature of the strata penetrated in boreholes without coring, allowing faster 'openhole' (i.e. no coring) drilling methods to be used for most of the strata. However, the coal seams are usually cored because samples required for chemical analysis. Sometimes part of the coal core is either broken or completely ground away during drilling and the geophysical logs can then provide a valuable check on the amount of core recovered and hence the seam thickness.

A string of electronic instruments is lowered down the borehold to log the strata after drilling has been completed. Several properties are measured during *geophysical logging*, but the most useful are gamma radiation and density.

Natural gamma ray emission from radioactive isotopes of potassium and uranium can be used to distinguish between shales, siltstones and sandstones. Potassium is present as a constituent of the mineral mica in most shales and hence measurement of its emission provides an evaluation of the clay content of rocks in the borehole. Marine bands frequently show high radioactivity owing to traces of uranium.

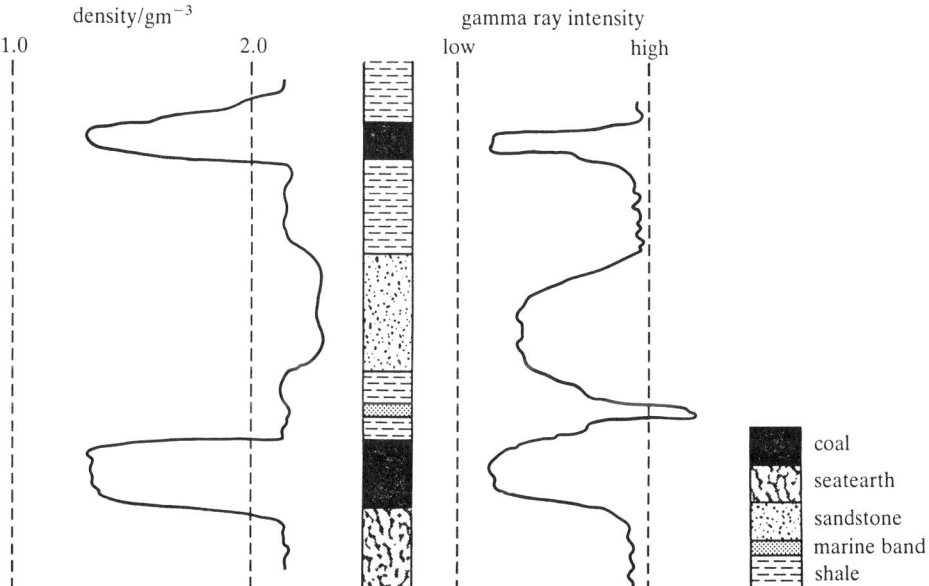

Figure 19   A geophysical log showing the density and gamma ray intensity data for a coal-bearing sequence.

Coal has a density of less than $1.4\,\mathrm{g\,m}^{-3}$ and the non-coal strata have densities of over $2.0\,\mathrm{g\,m}^{-3}$, so differences in density provide a good method of distinguishing coal from other rocks.

4.2.2 Seismic reflection profiling

Seismic reflection profiling is another geophysical technique adapted from the petroleum industry and used to identify structures, principally faulting, in coal sequences. Measurements of natural seismic waves generated by earthquakes are used to decipher the internal structure of the Earth; artificially generated seismic events form the basis of this geophysical technique. The times taken for seismic waves to travel from the source to surface detectors via reflecting interfaces are processed to produce profiles that give a picture of the geological structures at depth. Because of its low density coal has very different seismic characteristics from the surrounding rocks and produces strong reflections.

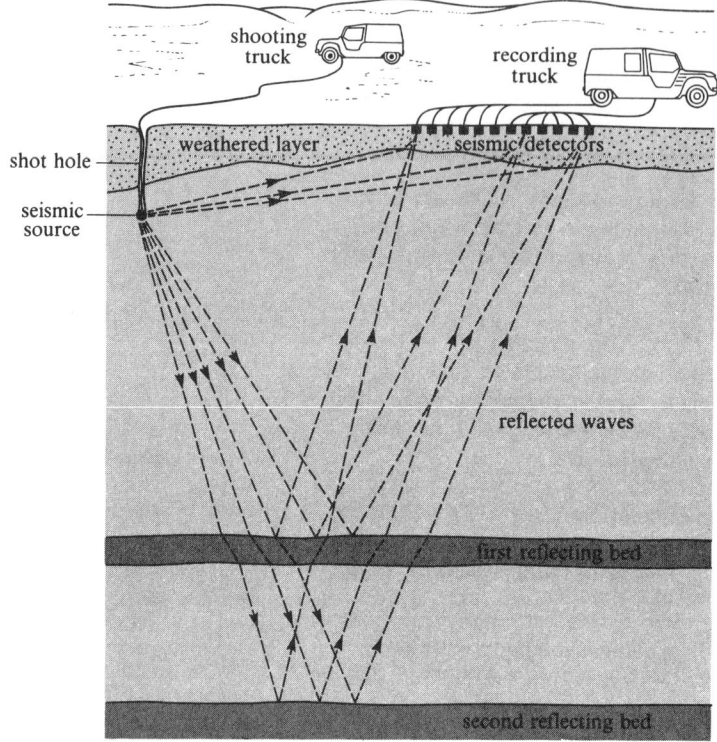

Figure 20 The layout of a seismic survey, showing the reflection and refraction of seismic waves by two reflecting beds.

The three principal components in a seismic reflection survey, as illustrated in Figure 20, are:

1   An artificial source provides the seismic waves: a small charge of dynamite is the usual source, but variants are used, such as a weighted pad vibrated hydraulically — a technique commercially known as 'vibroseis';
2   A series of detectors to detect the arrival of the reflected waves at the surface;
3   Recording equipment to record the electrical signals digitally on a magnetic tape.

Figure 21 is a reflection profile provided by British Coal. It shows a series of light and dark lines which approximate to the orientation of the sedimentary strata at depth. The intensity of the lines is related to the energy reflected from different layers. Organic materials such as coal and oil have low densities and low seismic velocities relative to the surounding sedimentary rocks. This means that they strongly reflect seismic waves, producing intense black or white lines in a seismic profile relative to the weaker lines produced by other sedimentary layers.

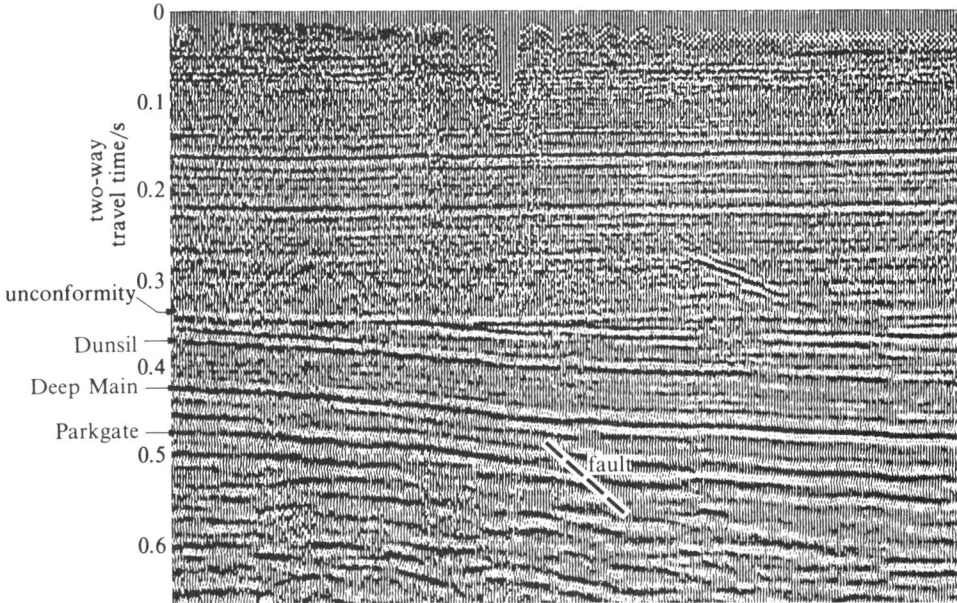

Figure 21 A seismic reflection profile showing a sequence of horizontal Triassic and Jurassic strata separated from gently dipping Coal Measures by an unconformity. The locations of the unconformity, three important coal seams (Dunsil, Deep Hard and Parkgate) and a fault are indicated.

4.2.3 Exploration in Britain

You will recall that exploration essentially involves focusing attention on the best of several prospects and that the evaluation of the results at each stage of exploration will shape and modify the plans for the succeeding stage. This 'evolutionary' approach to coal exploration is summarized in Table 3, in which six stages are defined.

Exploration for new mines will progress through these stages with the objectives of controlling exploration effort and bringing forward the best options first. Consequently the various prospects studied in Britain in recent years are now at different stages of exploration. Some have failed and some are on the shelf while others are being actively explored, are at the planning stage, or under development.

Most of the prospects examined in Britain by British Coal since 1970 are shown on Figure 22 and are briefly discussed below.*

In Scotland the exploration of the Hirst basin in the Upper Forth Valley has been the most successful and a mine area has been established with a seam

---

* Information taken from a paper by R Goosens, Colliery Guardian, August 1985.

around 2m thick and 400–600m deep. This mine can be worked as an extension to the large Longannet mine.

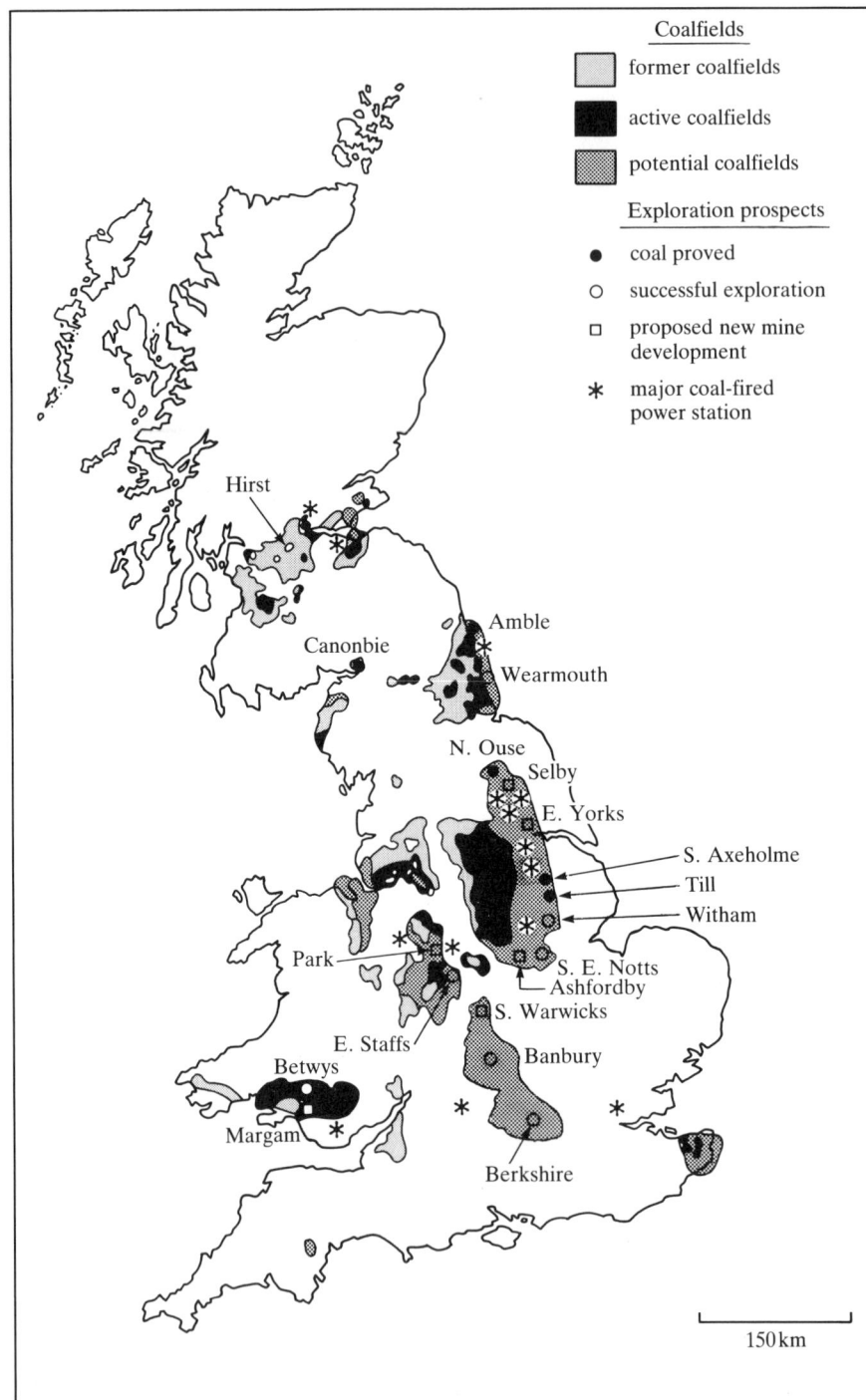

Figure 22 The locations of the British coalfields and the principal coal-fired power stations. The plan shows proposed new mine developments and exploration areas in 1984.

Table 3 An evolutionary approach to coal exploration

	Ladder of exploration knowledge	Degree of refinement at each rung of the ladder
	6 Development stage	a planned mining project being developed into a production unit
	5 Planning stage	feasibility studies successful and mine planning in progress
	4 Feasibility study stage	exploration results justify feasibility and design studies for a mining project
	3 Intensive exploration stage	areas with defined boundaries justifying intensive exploration
	2 Preliminary exploration stage	areas with coal but boundaries uncertain
	1 Potential prospect stage	areas where coal is believed to exist

(Left-hand vertical labels: increasing economic prospects ↑ · decreasing geological uncertainties · increasing costs)

Off-shore exploration programmes using a drill ship for the North East England coalfield have revealed a northerly extension east of Amble in Northumberland but access will be *difficult* due to faulting and folding and the fact that the reserves lie over 2 km from the coast. Further south exploration off the Durham coast has proved major extensions for existing collieries.

In Yorkshire the past 20 years have seen new *reserves* proved over a larger area than had ever been thought possible. In the north the coalfield extends to the north of the city of York with the North Ouse prospect containing thick coal seams but with more than average faulting. South of Selby the East Yorks prospect has located several good seams at depths of 600–1000m. However the real prize in Yorkshire has been the proving of the Selby coalfield. Planning consent to develop the Selby mine complex was given in 1976, twelve years after the initial boreholes proved the first workable coals. Some 60 boreholes and over 450 km of surface seismic line have provided the information upon which the planning and development of the project was based. The Selby complex consists of a surface drift mine for coal production and five satellite shafts for men and materials and is planned to produce 10 million tonnes per day when full production is reached.

Exploration in the Nottinghamshire coalfield has provided several attractive prospects to the east and south east of the established coalfield, while in the south much exploration has been completed in the NE Leicestershire prospect with planning consent granted for a new mine at Ashfordby in the Vale of Belvoir.

An unexpected extention of the Warwickshire Thick Seam has produced the South Warwick prospect near Coventry where design studies for a new mine are in progress. Exploration in East Staffordshire has established two prospects but seam quality problems make marketing the coal difficult and so these projects are at the feasibility stage. Finally in the South Midlands an increasingly attractive prospect is emerging from exploration in Oxfordshire and Berkshire, indicating that new coalfields may be located outside the traditional mining areas.

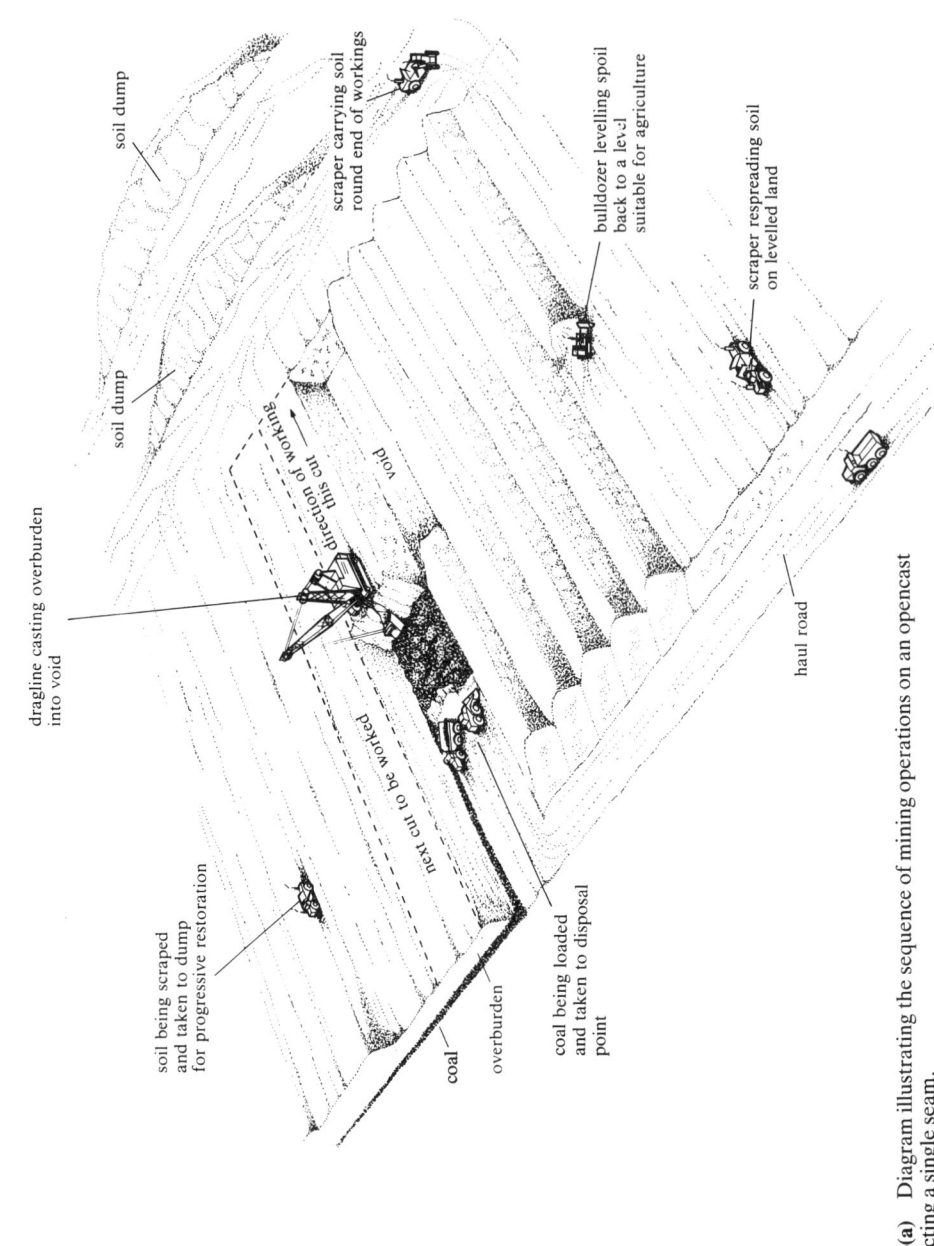

soil dump

soil dump

scraper carrying soil
round end of workings

bulldozer levelling spoil
back to a level
suitable for agriculture

scraper respreading soil
on levelled land

dragline casting overburden
into void

direction of working

void

next cut to be worked

haul road

soil being scraped
and taken to dump
for progressive restoration

coal

overburden

coal being loaded
and taken to disposal
point

Figure 23(a) Diagram illustrating the sequence of mining operations on an opencast site extracting a single seam.

While South Wales is geologically a difficult coalfield due to severe tectonic disturbances the value of the coking coal and anthracite reserves are very high and so exploration is continuing to locate new reserves.

4.3 Surface mining

The conventional view of coal mining is one of miners working deep in the earth in cramped and dangerous conditions. While most of Britain's coal is still extracted by underground mining methods, recent developments in surface mining have made this branch of the mining industry the most profitable part of British Coal's operations. Overseas, surface mining is providing most of the coal from the new coalfields developed in the 1960s and 1970s.

Opencast mining (sometimes known as strip mining) is the usual method of surface coal mining where the waste is dumped inside the worked-out section of the excavations. Figure 23a illustrates the method of opencast mining on a site working a single seam while Figure 23b provides a view of an opencast site in Northern England.

Figure 23(b) Opencast mining on Butterwell Site, Northumberland. A large dragline is removing overburden from the lowest seam mined while coaling and overburden removal are proceeding on higher benches to the left of the dragline.

Where several seams are present on a site, they are excavated on separate *benches* with the overlying strata or *overburden* removed by power shovel and trucks and a *dragline* casting overburden into the void of the deepest seam. The top surface of coal exposed on each bench is carefully cleaned to remove any adhering mudstone and is then excavated by hydraulic digger and loaded into trucks for despatch from the site. This careful exclusion of non-coal material, the 'dirt' or 'stone' impurities, enables consistently high quality coal to be produced by open-cast mining.

In Britain, open-cast sites are located in areas of shallow coal where underground mining has ceased and may include highly faulted ground or districts where the old mining methods had extracted only a proportion of the coal. Working depths of 75 metres are commonplace, with the deepest excavation of 260 metres at Westfield in Scotland. The maximum economic *stripping ratio* of overburden to recoverable coal has steadily increased over the years, as the value of the coal and the productivity of the equipment has improved, to levels of around 20:1 today, with the highest economic ratio of 35:1 being attained at a high value anthracite pit in south Wales. Open-cast mining is less costly and more flexible than underground mining methods and recovers a high proportion of the coal in the strata, up to 90 per cent, taking seams as thin as 0.15 metre.

4.4 Underground mining

About 90 per cent of Britain's coal is extracted by underground mining and highly mechanized extraction techniques were developed during the 1960s and 1970s to improve productivity. The coal seams exploited in most British mines, and those of Western Europe as well, vary from 0.5 to 3 metres and are typically about 1.5 metres thick; the Barnsley seam at the Selby mine is outstanding, with a thickness varying from 1.8 to 3.3 metres and it is an exceptionally good proposition for British mines. Underground mining is much less flexible and more labour intensive than surface mining. The design of an underground mine depends upon three factors; how best to plan the coal extraction system; how to ensure that a minimum of waste is mined; how to ensure safe operations.

Almost all the output from British and European underground mines is now produced by *longwall* methods, and a modern *coalface* is a production line between 50 and 250 metres in length with supply and ventilation tunnels at either end linking it to the colliery's communication system.

The likelihood of encountering face-stopping geological disturbances, such as faults, is one factor considered in the design of the coalface layout. From a resource aspect it would seem prudent to extract all the coal in a seam but in practical terms this can never be achieved. Roof or floor coal may be left to provide better roof or floor conditions; faulted areas may not be worked; support pillars will be left to prevent roof collapse and subsequent damage to underground roadways or surface buildings.

Figure 24 illustrates two common types of underground layout used in British mines. Figure 24a shows an *advance face* layout where the coalfaces are advancing into a block of coal, with the access roads at either end of the faces being advanced in time with the faces. Figure 24b illustrates a *retreat face* layout, in which the access roads are first driven to the far boundary of the block of coal before the coalface is opened at their extremities.

The coal is extracted by one or more shearer/loader machines which are propelled to and fro along the coalface, cutting and loading the coal directly onto a conveyor. The *shearer* consists of a drum fitted with picks rotating on a horizontal axis. It rotates at high speed, cutting into the coal up to depths of a metre as it advances along the face. The roof is held up by hydraulically powered supports which, together with the face conveyor, are advanced by hydraulic rams after the shearer has passed along the face. The roof behind the supports is then left unsupported and collapses. Such a face will be staffed by as few as seven men each shift and is capable of producing over 5000 tonnes of coal each day, ground to small sizes and well suited to the needs of power stations.

Figure 24c shows a ranging drum shearer operating on a coalface. A modern coalface is a very complex mining system and represents a large capital investment, up to £4 million for each face (1985 prices). Two consequences arise from the assessment:

1   The high capital cost of equipping a coalface means that a considerble period of trouble-free operations is necessary to recoup the financial investment.

2   This mining system is very inflexible, requiring uniform conditions to maximize its potential, and is incapable of negotiating any serious variations in the thickness of the seam.

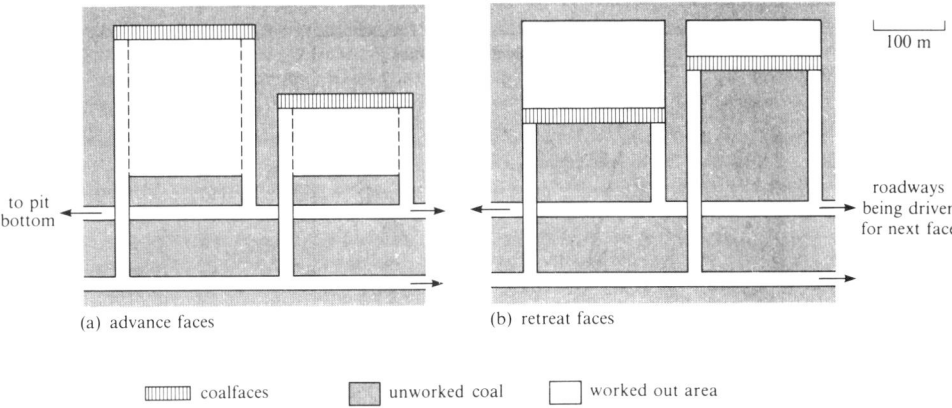

to pit
bottom

(a) advance faces

(b) retreat faces

100 m

roadways
being driver
for next face

| [IIIIIII] coalfaces | [▒] unworked coal | [ ] worked out area |

Figure 24 Two types of coalface layout: (a) advance face, where the coalface is advancing into a block of coal; (b) retreat face, in which the access roads are first driven to the boundary before the coalface is opened at their extremities; (c) 200 HP ranging drum shearer on the cutting run 20m from the main roadway.

What then are the effects of geological variations on such a complex mining system?

Geological factors are the primary controls governing the selection of working areas, the continuity of face operations and hence profitability. There are two principal categories of geological conditions which affect mining operations, the first relating to the structural setting of the seam (inclination or dip of the strata, presence of faults), and the second to its sedimentary setting (nature of the strata, lateral variations in rock type).

4.4.1 Structural setting

Folding in most coalfields is not sufficiently severe to impose limits on mine planning, but faults form major hazards for mechanized faces, as any displacement greater than seam thickness may bring production to a halt with consequent loss of output and revenue. We noted earlier that all British coalfields are subject to faulting to various degrees, the most severely disturbed areas being in south Wales.

4.4.2 Sedimentary setting

Whereas faulting may impose physical constraints on planning the layout of mine workings, the sedimentary setting of a seam and its associated strata determines the profitability of the mining operations. Variations may occur both within the seam and in the strata forming its roof and floor.

The thickness of a seam has a considerable bearing on profitability with the 1–2 metre thickness range particularly suited to mechanized longwall mining because the roof support equipment is most effective in this range. Seams less than a metre thick suffer from the law of diminishing returns as increasing areas of extraction (and hence more rapid face advance) are required to obtain the same bulk output. The presence of shale layers within a seam section introduces inferior material into the product and results in reduced proceeds, because the marketable value of coal is determined by its quality as measured by ash content. Such layers may indicate the beginning of a split within the seam.

Often of equal importance in achieving rapid face advance is the behaviour of the roof and floor strata. Shales and siltstones usually provide good working roofs for a coalface, but a soft seatearth may have a deleterious effect on the coal quality because the heavy equipment may gouge out several centimetres of floor material with the coal.

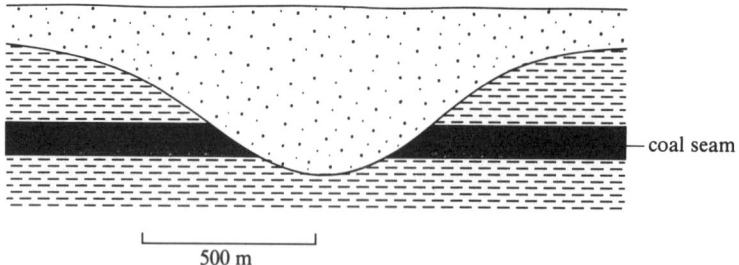

500 m

Figure 25 Diagrammatic cross-section of a washout.

Sandstones occur close to coal seams less frequently than the other rock types (Figure 7) but their presence can result in operational problems on the coalface. The most serious problems arise where *channel-fill deposits* are encountered in the shales above a coal seam. These structures represent erosive drainage channels on the original delta plain which had cut down into the underlying sediments and were filled with coarser sediments. When such channels merely cut down into the shales above a seam, they result in unstable roof conditions where serious roof falls may occur. However, sometimes the channels cut down into the coal seam and, in extreme examples, may cut it out altogether. Such structures are known as *washouts* (Figure 25) and they will bring a face to a standstill.

4.4.3   Recognition of geological hazards

Geological hazards can be classified into two types — gradual changes and sudden changes. Where a change is gradual, as with seam thinning or splitting, data from boreholes in advance of the workings supplemented by measurements taken on working faces and development roads can be used in the preparation of *isopachyte* plans. These plans enable the geologist to delineate those areas where a seam becomes too thin or split for workings to proceed.

Where the changes are sudden, as with faults or washouts, such detection techniques will only have limited success and other methods have to be employed. For example, detailed studies of the strata exposed in roadways and in borehole sections have enabled the paths of distributary channels to be plotted. The forecasting of fault positions is more difficult and seismic reflection profiling has been developed to locate major faults. Since 1980 a novel form of seismic exploration has been developed to detect disturbances within the seam at distances of a few hundred metres from mine roadways and coalfaces. This technique of *in-seam seismics* makes use of seismic charges to generate a wave which travels along coal seams and is reflected by any sudden changes, such as a fault or washout. It has been applied in many collieries with a high degree of success.

On an advance face the information about the geological conditions that lie ahead will be limited to that which can be deduced from any nearby workings or boreholes. So an advance face is a high risk layout, to be used in areas where uniform conditions are anticipated.

On a retreat face the access roads expose any geological troubles in the block of coal to be extracted, and the face can be scheduled to work back towards the main roadways without risk of interruptions. About one quarter of the output from Britain's mines in 1983 came from retreat faces, with their average daily output some 24 per cent higher than that from advance faces. Productivity is much higher on retreat faces because any geological hazards have been located as the access roads were driven. Advance faces are always liable to encounter unexpected bad conditions which reduce production or even halt the face.

4.4.4   Coal extraction

The proportion of the coal present in the ground that can be extracted is known as the *recovery factor* and varies considerably between collieries, ranging between the extremes of 73 and 24 per cent. In the new Selby mine, for instance, only 37 per cent of the reserves in the Barnsley Seam were classed as recoverable, largely because of the need to leave much coal in the ground in order to minimise subsidence.

Nationally, only about 50 per cent of coal is recovered from the working areas in underground mines. In addition, other areas of coal are excluded from the working areas because they are too thin or too disturbed by faulting or for many other reasons. consequently, the proportion of the total coal in situ that can be extracted is considerably less than 50 per cent.

When we consider the tonnage of coal mined, we must also remember that there is a variable tonnage of 'dirt' extracted in the *raw coal*. The early miners used to leave the small coal and dirt undergound, only loading the large coal. Modern mechanized mining cannot be so selective and large amounts of dirt are often brought out of the mine along with the coal. The raw coal emerges at the surface with a considerable proportion crushed to small sizes (less than 5 cm) during mining and transportation. For instance, in 1979 the British Coal North Yorkshire Area produced 8.2×10^6 tonnes of coal and 6.1×10^6 tonnes of dirt, which had to be tipped. Although a lot of this spoil was rock excavated during tunnelling, much also came out with the coal. The proportion of clean or saleable coal recovered from the raw coal output is commonly 60–70 per cent, but may be significantly less.

4.4.5 Processing the coal

Parallel with the developments in mechanized mining, coal preparation plants have been developed to sort the product on the surface. The most important plant is the *coal washery* which uses techniques developed from metalliferous mining. The principal equipment is the *jig* washer, which uses water to separate the larger sizes of coal from dirt by virtue of their differences in density. A jig washer is a water-filled box in which the particles are supported on a screen, illustrated in Figure 26a. When water is pulsed up and down through the screen and the overlying bed of particles, the particles separate into layers with the denser shale particles collecting towards the bottom, from where they are removed, and the lighter coal particles rising to the top for collection. The product then passes over *screens* to sort it into different size fractions.

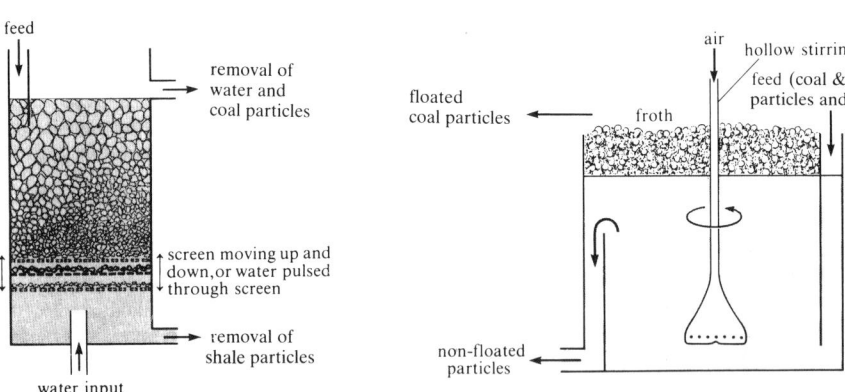

Figure 26a Diagram showing principles of jig washing.

Figure 26b Froth flotation cell.

Fine coal, below 0.5 mm, is cleaned in a froth flotation cell, Figure 26b. Special collector reagents are added to the slurry feed entering the cells which coat the coal particles in preference to the shale particles. Air bubbles rising through the cell adhere to the coal particles and carry them to the top of the froth liquid where they can be skimmed off while the shale particles are unaffected and pass through the cell in suspension.

One of the disadvantages of washing coal is that one inert (ash) is partly replaced by another (water from the washery) and so coal washing is kept to a minimum. Frequently different quality grades are blended to meet a customer's specification with a proportion of washed coal or clean opencast coal being blended with coal from the mine.

> Does open-cast coal require this elaborate treatment?

In Chapter 4.3 you learnt that the coal beds are carefully cleaned to remove any adhering shales. This enables consistently high quality coal to be produced without elaborate cleaning. However, it should be noted that the coal supplied to power stations does not have to be cleaned to produce low ash contents, but has to be prepared to the consistent standards as required by each power station. We will discuss this aspect later.

4.5 Environmental aspects of coal mining

You will be fully aware by now that the exploitation of mineral resources involves environmental issues as well as mining and geological ones. The importance of environmental issues in the coal industry was studied by a Government Standing Committee over the period 1978–81.

Many environmental issues are involved in open-cast mining, and proposals for new mines frequently arouse considerable local opposition. By its very nature open-cast mining has a major impact on the landscape, involving the digging of enormous pits with accompanyng noise, dust and traffic movements. However, sites have limited lives — up to 10 years — and engineers are justifiably proud of their record in restoring open-cast sites to productive farmland or to other purposes, such as forestry or recreational uses. Increasingly in recent years the environmentally conscious public has used legal rights in the planning processes to oppose and sometimes prevent mining on sites where its impact would be severe.

Deep mining operations have two significant areas of environmental impact — spoil heaps and subsidence. Spoil heaps have always been the principal surface feature of underground mining operations, but with the development of large mechanized collieries in the later twentieth century they have become increasingly intrusive on the landscape; since the Aberfan disaster, when a spoil heap collapsed and buried a school and houses in a south Wales valley, their siting and construction have been carefully controlled. Subsidence is an inevitable hazard in mining areas, with British Coal accepting liability to repair damage as part of the costs of coal extraction. The major factors affecting the extent of subsidence are the depth and the thickness of the seam extracted. The thicker the seam the greater the void left after mining, with consequently greater eventual subsidence at the surface. But the deeper the seam the more the effects of subsidence will be reduced through successive strata, so that the effects at the surface will be less although extending over a wider area. Certain vulnerable structures, such as dams, viaducts, and historical buildings, are protected by unworked pillars of coal left beneath them, but such protection may be extremely costly where it significantly affects the layout of the mine.

New mining schemes, such as the Selby project, are subjected to lengthy Public Inquiries and the approval to proceed with the project from the Secretary of State for the Environment may be subject to several stringent restrictions. At

Selby, these restrictions were related both to the visual impact of the surface developments at the mine sites and to minimizing the effects of ground subsidence in areas particularly liable to flooding or subsidence damage.

Throughout the country considerable expense and effort has been devoted in recent years to improving the visual appearance of mine sites and spoil heaps at existing collieries, by landscaping and by planting trees, and the situation has been considerably improved in many mining districts.

Another sensitive environmental issue is the subject of air pollution and acid rain which will be considered later in Part II Chapter 4.1.

4.6 Uses of coal

Coal consumption has declined drastically since 1950 when it comprised 90 per cent of the fuel consumed in Britain to a mere 35 per cent in 1982/3*. This decline is illustrated in Figure 27, which shows that the decline has affected all fuel users except the power station market where, in contrast, demand has virtually doubled.

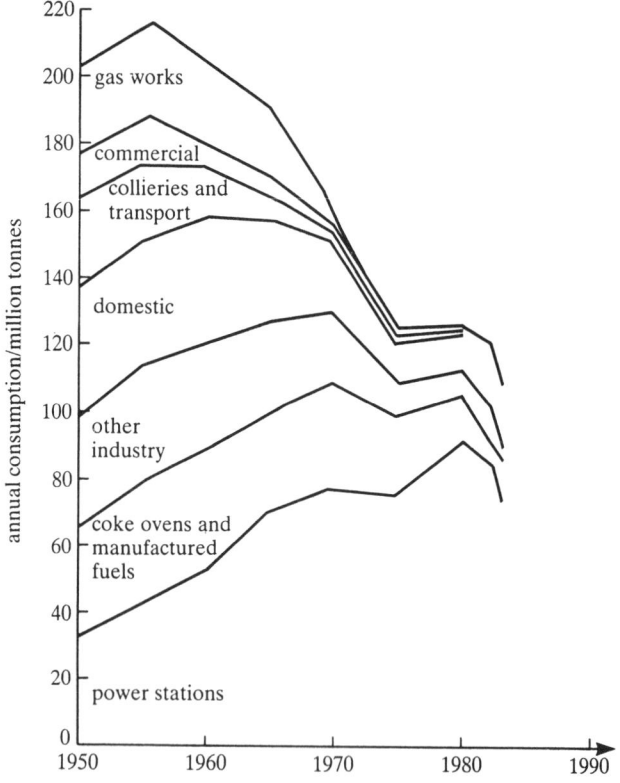

Figure 27 The changes in markets for coal in Britain between 1950 and 1983*.

Power Stations

Of the fuel used in 1983 for electricity generation, 75 per cent was provided by coal amounting to 86×10^6 tonnes (72 per cent of the total coal production; Figure 27). Clearly, British Coal and the Central Electricity Generating Board

* 1983 was the last year before industrial disputes (miners strike) affected production.

(CEGB) are highly dependent on each other. Power stations use the small grades (less than 5 cm) produced in large quantities by modern mining methods; the coals are first dried and pulverized to fine powder before being blown into the furnace by a stream of air as *pulverized fuel*. Practically any coal can be used, the most important characteristics being minimum costs and consistent quality, with usually about 16 per cent ash content, in order to achieve optimum operating conditions. Because the biggest part of electricity costs arises from the fuel charges, the electricity generating boards have built very large power stations in the most productive coalfields of Yorkshire and Nottinghamshire in England and the Forth Valley in Scotland. The location of the major coal-fired power stations is shown on Figure 22 together with the extent of the coalfields. We will come back to the subject of fuel for power stations in Part II.

Industrial uses of coal

Figure 27 shows the decline in the industrial uses of coal since 1950. Sales to this market in 1982/3 amounted to 9×10^6 tonnes, with energy-intensive industries such as brick-making and cement taking about a quarter of this tonnage. Industrial coal-burning appliances frequently require good quality, sized grades (1–5 cm) although some large consumers such as the cement works use pulverized fuel like the power stations.

Although coal costs in 1982 were very competitive with those of oil products, many industrial concerns were deterred from converting from oil by the higher capital costs of coal-fired equipment and the need to provide coal storage and ash disposal.

One new development, *fluidized bed combustion* holds out potential advantages over conventional equipment in that it enables a wider range of coal quality to be used and has considerable scope for reducing atmospheric pollution by absorbing sulphur dioxide in the ash. Coal is burnt in a bed of its own ash or other granular refractory material which is kept in a state of turbulent motion by the upward flow of combustion air through the bed. This new technique is competing with oil-fired industrial boilers and may find a role in the future design of coal-fired power stations.

Coke manufacture

This market has declined substantially since the 1960s to only 8×10^6 tonnes in 1982/3, owing to the run-down of the steel industry. Prime quality coking coal (Chapter 3.4) of low ash content is required, and the bulk of it comes from the south Wales coalfield. Since the 1950s much progress has been made in improving blast furnace techniques and this, together with the use of richer iron ores, had earlier led to considerable reductions in coke consumption. There are no indications of possible future expansion in this market and most coking coal in now imported.

Domestic uses

Coal sales to the domestic market have diminished rapidly owing to the introduction of clean air zones in urban areas and the development of natural gas supplies. Most domestic appliances use natural smokeless fuels (anthracite) or manufactured fuels (such as coke). These are the best quality fuels with low ash content (5–10 per cent) and are consequently expensive.

Chemical feedstocks

A new market that may emerge before the end of the century is the use of coal as a raw material for the manufacture of *synthetic natural gas*, known as SNG. This process involves the production of methane from coal by *hydrogenation* (i.e. increasing the ratio of hydrogen to carbon) and the product is very different from the old-fashioned town gas produced coke manufacture.

4.6.1 Future trends in coal use

It is notoriously difficult to forecast future energy demands and the plans of the coal industry have been over-optimistic since the early 1960s. In response to the transformed energy situation in 1974, a large capital investment was commenced to increase coal production and the initial "Plan for Coal" (1974) called for production to be raised to between 135×10^6 and 150×10^6 tonnes by 1985. These plans were reviewed in 1977 and it was then proposed that the annual production target for 2000 should be 170×10^6 tonnes (see Table 4).

Table 4 Present markets and future prospects for coal in Britain: four estimates of coal demand by the year 2000

Market	Usage in 1982/3[*] 10^6 tonnes	Estimates for year 2000/10^6 tonnes		
		British Coal[a]	Dept. of Energy[b]	Robinson[c]
power stations	90	90	78, 65	40–60
coke ovens	12	20	19, 12	13–15
industry	8	40	45, 32	15–25
domestic/commercial (inc. SNG)	18[†]	20	23,6	7–10
total	121	170	165, 115	75–110

† includes 7×10^6 tonnes exported
* last full year before industrial disputes

Notes to Table 4

(a) Estimates presented to the Commission on Energy and the Environment. Report of Government Standing Committee, 1981.

(b) Upper estimates (1979) based on 2.7 per cent growth rate; lower estimates (1981) based on 1 per cent growth rate.

(c) C. Robinson and E. Marshall (1981) *What future for British coal?* Hobart paper no.89, Institute for Economic Affairs, London.

1984 was a traumatic year for the British coal industry with the year long miners' strike. The dispute emphasised the importance of opening new reserves which can produce coal at a price that more than competes with overseas coal. In 1985 British Coal produced its "New Strategy for Coal", which marked the end of the indicative planning that has guided the industry's development since its creation in 1947. The new approach requires the industry to meet its own operating costs and to match the market prices set by foreign coal and the fuels. Such objectives can only be met by significant reductions in operating costs since international energy prices seem likely to remain highly competitive and market prospects look poor. At present British coal is produced at around £44 per tonne and British Coal aim to reduce costs to £42 per tonne in the short term and to

£39 per tonne within 5 years. In view of the inability of the coal industry to increase productivity and reduce production costs significantly in the past, it will be interesting to see how effectively this strategy can be applied.

It seems inevitable that coal consumption will continue to decline as more nuclear power comes on stream; the size of the British coal industry will depend upon imports remaining at not more than 8×10^6 tonnes and upon export sales increasing. An output of around 100×10^6 tonnes per year in 1990 seems likely, showing a marked decline of 20×10^6 tonnes from the immediate pre-strike production (Table 4). With a continuation of profitable coal production by opencast mining at 15×10^6 tonnes per year and the introduction of production from the planned new mines like Selby and Ashford by producing 20×16^6 tonnes per year, it seems probable that the existing collieries are likely to produce around $60–65 \times 10^6$ tonnes per year by 2000, compared with 105×10^6 tonnes in 1983. Thus the coal industry will continue to contract into a smaller number of very large production units increasingly concentrated in the Midlands coalfields with the consequent reduction of mining activities in the older mining areas of Scotland, NE England and S Wales.

4.7   British coal resources

The first attempt to asses Britain's coal resources was made in 1871 by a Royal Commission, which estimated 'total reserves' to be 148×10^9 tonnes. A survey in 1976 provided a gross figure of 190×10^9 tonnes of 'coal-in-place' remaining in Britain, defined as material containing less than 40 per cent ash, in seams at least 0.6 metre thick and less than 1220 metres below the surface. This value is only of academic interest and it is a gross over-estimate of economically workable reserves.

More important are estimates of economically workable reserves because they will be the basis for decisions on future energy policy and investment in the coal industry. Thus in 1946 the Ministry of Power reported a total of about 36×10^9 tonnes of 'workable coal reserves' in seams at least 0.6 metre thick and less than 1220 metres below the surface. An estimate in 1976 was that 45×10^9 tonnes of coal were considered to be 'recoverable and extractable under present conditions'. At current rates of extraction that would last for 300 years. This estimate has been strongly criticized as being over-optimistic and based on deduced rather than proved *reserves*. Since very few decisions need to be taken based on such long term forecasts, the actual tonnage is of little significance and these estimates cannot be considered as reserves but should be classed *conditional resources*.

The mining engineer and the geologist look at coal reserves over an extended timespan to answer questions concerning present and future mining capacity. Much more rigorous definitions of 'workable' reserves are needed for these purposes. For an older colliery with little coal left, the estimates may relate to rather poor quality reserves for the next 5 to 10 years production, whereas for a new mine with very large amounts of coal in the ground, such as Selby, reserves estimates may relate to the next 50 years production and still exclude whole seams because they have not yet been proved, appear to be less attractive or simply are not needed. These reserves were classed by British Coal in 1985 as 'operating reserves' and amount to a total of 5×10^9 tonnes for Britain. Even this more rigorous estimate means that there is sufficient coal to last until well into the twenty-first century.

4.8 Summary of Chapter 4

1   The British coalfields are of Carboniferous age and were deposited in vast deltaic swamps which covered much of Britain and the southern North Sea. The Carboniferous sediments of southern England and south Wales were severely deformed in the Variscan orogeny. The more northerly areas of Britain experienced less intense deformation, and the resulting uplift and erosion separated these areas into isolated basins before the deposition of Permian and Triassic sediments.

2   Coalfields are either exposed or concealed, depending whether the Coal Measures rocks are hidden by younger strata. In most coalfields, mining commenced in the shallower exposed regions and gradually extended into the deeper parts of the concealed coalfields.

3   The most effective and widely used exploration technique is drilling boreholes to provide solid cores of coal seams for chemical analyses. Geophysical logging records of the strata cut in the boreholes without coring, while seismic reflection profiling provides a continuous survey of major fault locations between boreholes and other structural information.

4   Most of Britain's coal is extracted by underground mining, usually using the longwall method with mechanized systems for coal cutting, coal transportation and roof support. Such systems represent a very large capital investment and require considerable periods of trouble-free operation to recoup their costs. They are also very inflexible mining systems and are incapable of negotiating many of the geological variations likely to be encountered. Consequently an understanding of the geological variations likely to be met is essential both in selecting working areas and in maintaining the continuity of face operations.

5   Open-cast mining results in short-term environmental disturbances; underground mining provides lasting problems, owing to spoil heaps and subsidence.

6   There are currently four principal markets for coal in Britain, of which the power stations are by far the largest. The others are industrial and domestic consumers and coke manufacture. Each of these markets has special requirements.

7   Estimates of the future requirements for coal show considerable differences. However, exploration has confirmed that Britain has very extensive coal resources.

5   The world coal economy

Study comment. First we review the world coal trade, highlighting the predominance of a small number of producing countries. Finally we examine the extent of the world's coal resources.

5.1   International trade in coal

The markets for coal in Britain discussed in Section 4.6 are not necessarily identical with the demand for British coal; both the CEGB and the British Steel Corporation have attempted to import coal from overseas suppliers but have been restrained by governments in order to protect the markets for British coal. In contrast, on the international market British coal has to compete with other supplies and has achieved little success in this direction. During the late 1970s and early 1980s annual exports of British coal barely exceeded 2×10^6 tonnes, although they did increase in 1982 to 7×10^6 tonnes, while the markets that this coal could most readily serve — countries of the European Economic Community (EEC) — have been increasing their coal imports steadily during recent years, from 30×10^6 tonnes in 1973 to 75×10^6 tonnes in 1980.

Table 5 compares the cost of surface and deep mine coal in various producing countries and shows that overseas surface coal is considerably cheaper than west European deep mine coal.

Table 5 Comparative international costs for mining coal in 1979[*]; in US dollars per tonne

Producing country	Underground	Surface
UK	45–135	average 38
West Germany	55	—
USA	20–35	8–18
Canada	—	15–20
Australia	15–25	12–20
South Africa	10–15	8–10

[*]European data derived from NCB sources and other data from World Coal Study, 1980. (Confidential data in 1983 showed comparable prices for the different sources.)

It is an inescapable fact that the British coal industry, like those of other west European nations, is primarily a deep-mine industry with only some 15 per cent obtained by surface mining, in contrast to a world coal industry which largely uses surface mining technology with its low labour costs. Half the world coal output now comes from surface mines, particularly in the exporting countries, and the proportion is expected to increase steadily. The total world coal production in 1983 is listed in Table 6 and the tonnages exported by the principal coal trading nations are also given.

Table 6   World Coal Production in 1983

Country	Bituminous coal and anthracite 10^6 tonnes		Lignite 10^6 tonnes
Developed Countries	*Production*	*Exports*	
W Europe	246	7[*]	186
USSR	558	–	158
E Europe less USSR	232	18[**]	572
USA	712[†]	55	[+]
Canada	37	17	8
Australia	101	58	35
Developing Countries			
P R China	715[†]	–	[+]
India	136	–	7
Asia less China & India	92[††]	[††]	28[†]
Latin America	22	–	–
Africa	152[‡]	30[‡]	–
World Total	3003		994

Notes

[*]   Exports from UK
[**]   Exports from Poland
[†]   Includes production of hard coal and lignite
[††]   Japan only produced 17×10^6 t but imported a total of 89×10^6 t, principally from Canada and Australia
[+]   Includes 21×10^6 t from Turkey
[‡]   Includes 146×10^6 t from S Africa, which country exported 30×10^6 t

There are six countries with annual exports of more than five million tonnes, of which the USA and Australia are the principal exporters, while there are 23 countries with annual imports exceeding a million tonnes, with Japan being the largest single importer.

The USA with its vast coal reserves could become the 'Saudi Arabia' of the world coal trade and is steadily increasing its coal exports to western Europe and the Far East. Australia, China, Canada and now Columbia* are all competing to fuel Japan's rush to convert its power stations to coal burning. Australia enjoys the advantage of having its important mining centres in New South Wales and Queensland close to ports and Australian coal is highly competitive in most parts of the world, including western Europe.

New coalfields are being developed in various parts of the world, both by national governments and by multinational companies. One important commer-

* One of the largest surface coal mines in the world is being developed at Cerrejon, Columbia, by the oil company Exxon, due to produce 15×10^6 tonnes by 1989

cial difference between the exploitation of coal and oil, however, is that no international producers group or cartel for coal is likely to emerge, in contrast to OPEC in the oil trade.

5.2   World coal resources

One of the major problems in any estimation of world coal resources is the very considerable differences in the states of geological knowledge in different countries, for example the USA and Brazil. In the USA very extensive geological investigations and drilling programmes have located probably all the major coalfields and now attention is directed to extending knowledge in these fields. However, the remote and undeveloped character of much of Brazil means that there may be important coalfields awaiting discovery. Their distance from industrial centres or ports and the lack of transportation prevents any real interest in developing such regions.

The world's energy resources are reviewed periodically by the World Energy Conference (WEC). The statistics produced by WEC are always complicated by the different interpretations used in the various countries of the criteria for assessing reserves, sometimes influenced by political and economic interests.

WEC defined various standards to minimize these difficulties and used the following definitions:

Hard coal includes bituminous and anthracitic coals

Brown coal includes sub-bituminous and lignitic coals

Recoverable reserves reserves that are actually recoverable under the prevailing technical and economic conditions, including:

	hard coal	brown coal
maximum depth	1500 metres	600 metres
minimum seam thickness	0.6 metre	2 metres

Total resources   resources that may become of economic value at some time in the future, but with a constraint on the maximum depth of 2000 metres for hard coal and 1500 metres for brown coal.

The results of the 1983 WEC appraisal of world coal resources are summarized in Table 7. Figure 12 shows that coal deposits occur in a great many countries and in every continent. However, ten countries account for about 92 per cent of currently estimated coal resources and almost 90 per cent of the reserves. Moreover, three countries — USSR, USA and China — contain 83 per cent of the world's coal resources and 60 per cent of the reserves. The magnitude of the world's coal resources is difficult to comprehend fully because the tonnages are so vast in relation to current and future markets for coal. It is clear therefore that the world's coal resources are well able to support any demands that are likely to develop within the foreseeable future. The biggest problem is likely to be the vast scale of mining investment and development which would be required to utilise their potential. We will return to this theme when we consider the roles of the different energy sources.

Table 7 World coal reserves and resources (1983 revision)

Region	Total resources /10⁹ tonnes	Recoverable reserves /10⁹ tonnes
N America	2 685	187
S America	35*	10*
W Europe	419	82
E Europe	170	46
USSR	4 860	110
China	1 438	99
India	57	34
S Africa	173	34
Australia	263	27
World totals	10 100	629

*Low estimate.

5.3   Summary of Chapter 5

1   Despite the increasing demand for coal imports by the EEC countries, Britain has made little contribution to the supply because the British coal industry is primarily a high cost, deep-mine industry. Half the world's output now comes from surface mines, which produce coal considerably more cheaply than the western European deep mines. There are six countries with annual exports greater than 5 million tonnes, of which USA is the principal exporter, while 23 countries import more than a million tonnes with Japan as the largest single importer.

2   The world's energy resources are reviewed periodically by the World Energy Conference, and their appraisal indicated that there are vast resources of coal available in many countries, with ten countries accounting for the bulk of the recoverable reserves.

6 The nature, origin and generation of petroleum

Study Comment. In this Chapter we consider the composition and processes of formation of petroleum which, after water, is the second most abundant fluid in the Earth's crust. The conversion of organic matter into petroleum is compared with those processes leading to coalification.

6.1 Nature of petroleum

Petroleum is a general term for a mixture of various hydrocarbons, and the relative amounts of these compounds in a given sample determine its properties. Petroleum exists naturally in gaseous (*natural gas*), liquid (*crude oil*) and solid (*asphalt*) states. The commonest elements in the compounds of petroleum are thus carbon and hydrogen, with much smaller amounts of oxygen, nitrogen and sulphur (Table 8).

Table 8 Elemental composition of typical petroleum samples

Element	Crude oil (weight per cent)	Natural gas (weight per cent)	Asphalt (weight per cent)
carbon	82.2–87.1	65–80	74.4–80.2
hydrogen	11.7–14.7	1–25	7.5
oxygen	0.1–4.5	—	7.6
nitrogen	0.1–1.5	1–15	1.7
sulphur	0.1–5.5	trace–0.2	3.0

Petroleum normally occurs in sedimentary rocks deposited under marine conditions, and its complex nature is evident from the fact that, so far, over 1 200 different hydrocarbons have been identified in crude oil. The majority of these compounds contain between one and 40 carbon atoms, and the number of carbon atoms is the basis of a means of grouping the components of petroleum. The seven main groups, or *petroleum fractions*, are gases (one to three carbon atoms), gasoline (C_4 to C_{10}), kerosine (C_{11} to C_{13}) diesel fuel (C_{14} to C_{18}), heavy gas oil (C_{19} to C_{25}), lubricating oil (C_{26} to C_{40}) and waxes (over C_{40}).

So petroleum, like coal,is a carbon-rich material, and there are also similarities in their processes of formation. There is one crucial difference between them, however. Coal is solid and stays where it was formed; petroleum in its fluid forms can move away from its source and accumulate in favourable structures elsewhere in the sedimentary sequence.

6.2 Origin of petroleum

There is no doubt that most of the compounds that have been identified in petroleum are organic in origin. Ideally, the organic matter is deposited on the sea floor, where it must be protected from scavengers that would consume and destroy it. This protection is afforded by (i) anaerobic conditions, which arise because the bottom waters are free of oxygen, and (ii) a relatively high rate of sedimentation of inorganic material, such as clay or sand particles, to bury the organic matter.

Table 9 lists the annual production of organic carbon from each of the seven main types of environment found on Earth. Note that terrestrial plants produce a total of 5×10^{10} tonnes yr^{-1} and that the marine environments also produce 5×10^{10} tonnes yr^{-1} of organic carbon. The continental shelf is the most productive region of the marine environments; while it has an area of only about one third of that of the open ocean, it produces more organic matter.

The continental shelf regions are more productive than the open oceans because of the higher nutrient content of shallow water; there is an added consideration because rivers transport large amounts of land-derived carbon into shallow seas. So present-day marine sediments can contain organic matter from several sources.

Table 9

Environment	Area/10^6 km^2	Organic carbon/ 10^{10} t yr^{-1}
desert	68	0.2
grassland	26	0.65
forest	41	3.25
agricultural land	14	0.90
near-shore seas	2	0.40
continental shelf	84	2.325
open ocean	276	2.275

However, you will recall that land plants appeared 400 Ma ago in the Devonian and that the first rapid development of these floras 345 Ma ago gave rise to the first vast deposits of coal in the Carboniferous (Chapter 3). So the main source of organic material before the Devonian must have been marine in origin. From the Precambrian until the Devonian, the largest producer of organic matter was phytoplankton. These are microscopic marine *algae* , which live in the upper layers of the oceans and on death sink in countless millions to the sea floor. There the algae become part of the sediment and form organic-rich marine shales. Figure 28 shows how the abundance of phytoplankton has changed with time. There were two peaks of maximum production: the first in the Ordovician — Silurian 500—400 Ma ago and the other in the Jurassic — Cretaceous 195—65 Ma ago.

Why did these peaks occur when they did, and is there a link between the phytoplankton abundance and crude oil reserves?

The two phytoplankton peaks tend to coincide with global rises in sea-level (Figure 28) and it has been suggested that, during the major rises, seawater flooded into lowland areas and into vast terrestrial depressions, where enormous phytoplankton populations developed in nutrient-rich marine basins. On death, these algae would sink into oxygen-free muds which on burial formed beds of organic-rich material. These contain 0.5 per cent or more of organic matter and have the potential to generate petroleum; for this reason they are termed *source rocks*. Source rocks are normally shales and contain about 90 per cent of all the organic matter found in sediments; on average, shales contain 1.2 per cent organic matter as compared with limestones and sandstones which have 0.56 and 0.05 per cent, respectively.

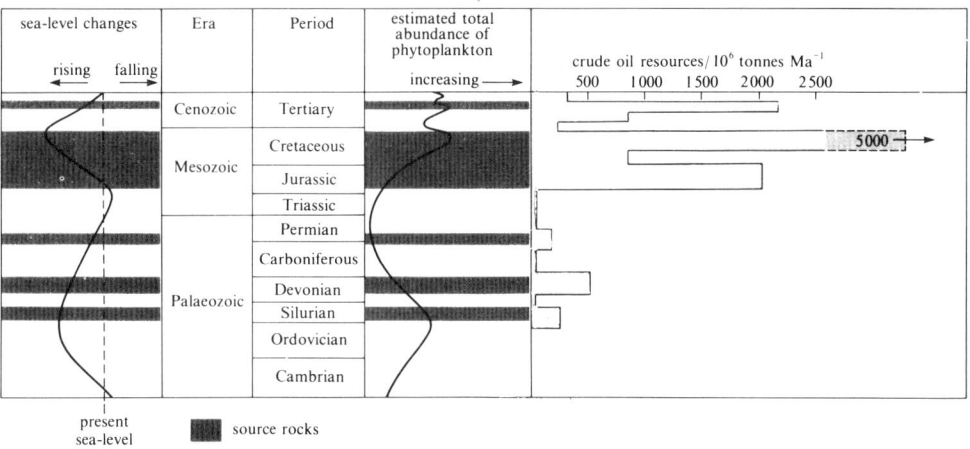

Figure 28 Major changes in sea-level and abundance of phytoplankton for the last 570 Ma. Also shown are the source rocks for petroleum and the crude oil reserves in relation to the stratigraphic column.

As you can see from Figure 28, the greatest volumes of source rocks are of Jurassic–Cretaceous age, and these are responsible for about 70 per cent of the world resources of petroleum. But how can the much lower level of Palaeozoic crude oil reserves be reconciled with phytoplankton production at that time? There is no great difference in the abundance of phytoplankton in the two Eras, so the explanation must be that the crude oil in the Palaeozoic rocks has been partly destroyed as a result of mountain building events, which caused deep burial, heating, folding and erosion of rocks; the Mesozoic rocks have not been so severely affected by these destructive geological processes mainly because they are much younger.

6.3 Generation of petroleum

The conversion of sedimentary organic matter into petroleum is termed *maturation* and can be divided into three phases (Figure 29).

1 *Diagenesis* Immediately after deposition, biochemical decomposition of the organic matter starts and produces 'biogenic' methane (cf. coal formation, Chapter 3.3). As the organic-rich sediment is buried and subjected to slightly increased pressures and temperatures the organic matter is converted into *kerogen*, a general term for a complex of rather amorphous material composed of carbon, hydrogen and oxygen that develops during the early stages of the maturation of marine organic matter.

2 *Catagenesis* With higher temperatures (and pressures) kerogen is altered and the majority of crude oil is formed. During this and the next phase of maturation (metagenesis), the larger molecules break down into simpler molecules; this process is called *cracking*, and is very similar to the process that takes place in oil refineries.

3 *Metagenesis* In the final stage of alteration of kerogen and crude oil, natural gas mainly in the form of methane is produced and residual carbon is left in the source rocks.

Although the main source of crude oil is phytoplankton, other types of organic matter such as land-plant remains, bacteria and zooplankton are present in marine sediments and can be converted into kerogen. Three kinds of kerogen from different sources have been identified, each of which yields different products during maturation. Table 10 shows how the hydrogen:carbon (H:C) and oxygen:carbon (O:C) ratios differ in the three kerogen types.

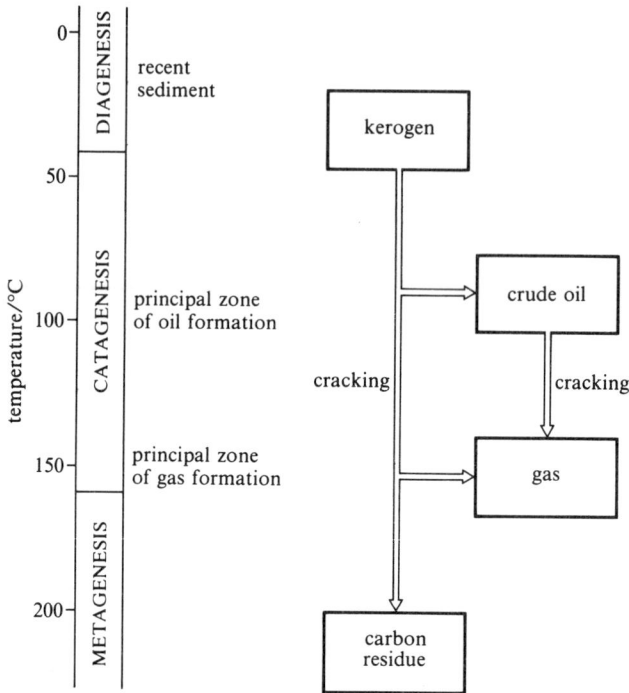

Figure 29 The products of the maturation of organic matter as the temperature increases during diagenesis, catagenesis and metagenesis; principal zones of oil and gas formation are shown.

Table 10 The characteristics of the three types of kerogen

Kerogen	Ratio H:C	Ratio O:C	Origin	Maturation products
type I	high	low	marine algal material	light high-quality oil (gasoline)
type II	high	high	mixture of various marine organic matter	main source of crude oil; some gas
type III	low	high	terrestrial	mainly gas with some oil; rich in wax

Although the maturation product is controlled by the composition of the original matter, the maturation of kerogen into crude oil and natural gas is achieved mainly by an increase in temperature, and therefore the formation of petroleum depends on the geothermal gradient in the crust. Figure 30 shows the relative proportions of crude oil and gas formed at increasing temperatures. Along the top are the numbers of carbon atoms in the chemical formulae of the various products. The higher number of carbon atoms represent the greater molecular weight of the organic compounds.

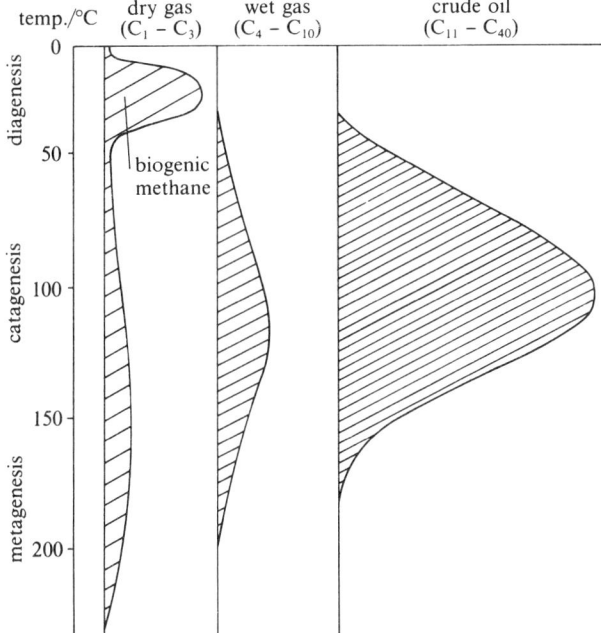

Figure 30 The temperatures of the maturation stages and the relative amounts of crude oil, wet gas and dry gas produced. The ranges of number of carbon atoms in each of the maturation products are also shown.

Petroleum is probably generated from kerogen in very small amounts at temperatures below 50°C with the peak production at about 100°C; when the temperature is raised to above 150°C even for a short time crude oil itself will begin to break down or 'crack' (Figure 29). The first gas to be produced contains some C_4-C_{10} compounds and is termed '*wet gas*'. But as the temperature increases these break down to give C_1–C_3 gaseous compounds and the product is then called '*dry gas*'. So the most important factor in the generation of petroleum is heat, which depends on the geothermal gradient in the sedimentary basins, and as you will see later in this Chapter such gradients vary from basin to basin. Source rocks that remain at very shallow depths will not normally generate petroleum: deeper burial, accompanied by an increase in temperature, is required.

The maturation pathways leading to the formation of petroleum are similar to those leading to the formation of coal but the end products differ because the starting materials are different. Figure 31 shows the maturation pathways for the three types of kerogen listed in Table 10; it also indicated how these relate to the maturation of coal macerals (Chapter 3.3 and Figure 11).

The lowest branch of Figure 31 is inertinite and there is no equivalent kerogen type. The next branch is kerogen Type 111 which Table 10 indicates is derived from terrestrial organic matter and which resembles vitrinite in composition and maturation. This kerogen type is not an important source of crude oil but can generate natural gas. Kerogen Type 11 is derived from mixed

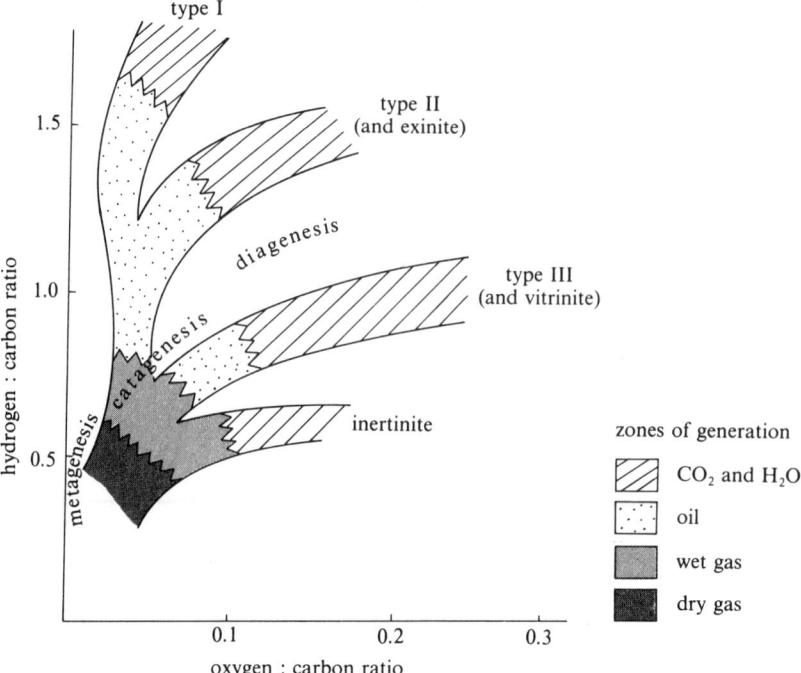

Figure 31 Maturation diagram for the three types of kerogen, showing the stages and products of maturation and the equivalent coal macerals.

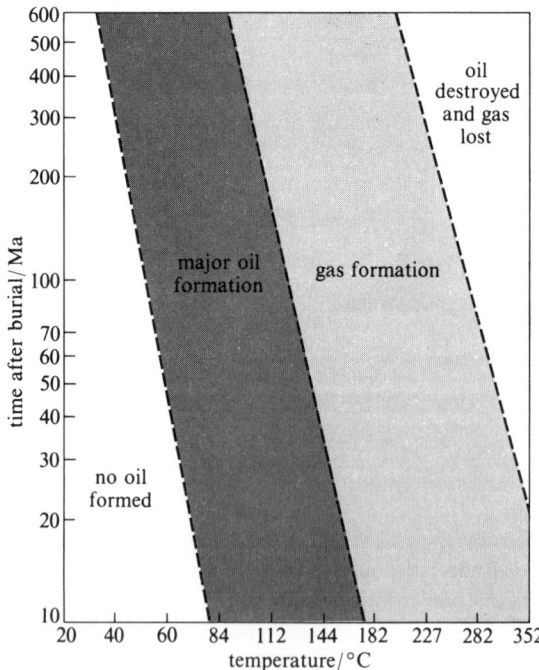

Figure 32 A model showing a possible relationship between the time after burial of source rocks and the temperature for oil and gas formation.

marine organic matter and is the main source of crude oil. It resembles exinite in composition and they both follow similar maturation paths. Kerogen Type 1 is the richest in terms of hydrogen to carbon ratio and has no terrestrial equivalent, being derived entirely from phytoplankton; it gives rise to a light crude oil.

Figure 31, like Figure 30, shows how maturation proceeds through the three stages of diagenesis, catagenesis and metagenesis. The areas representing the different hydrocarbon products are very approximately in proportion to the volumes of those products generated from the three kerogen types.

As with coal formation (Chapter 3.3), the time factor is also important in the generation of petroleum. It is assumed that different volumes of petroleum would be generated from similar source rocks if they were subjected to the same temperature for different lengths of time (e.g. Figure 32).

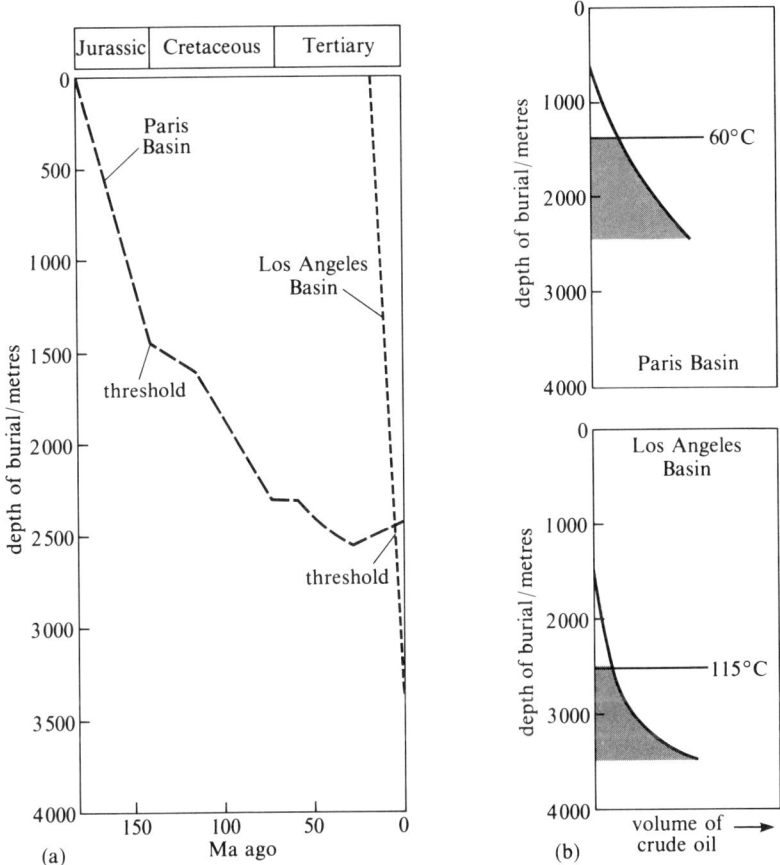

Figure 33 Reconstruction of burial histories of rocks from two basins. (a) The Jurassic Paris Basin rocks were buried to a depth of just below 2500 metres, whereas the much younger rocks of the late Tertiary Los Angeles Basin were buried to a depth of 3250 metres. (b) The principal zones of oil formation in relation to depth in the two basins and the estimated temperatures at which oil began to form are also shown.

This relationship can be illustrated by the following examples. Source rocks subjected to a temperature of 50°C for 30 Ma would not have been heated sufficiently to generate oil and only biogenic methane would be present while rocks heated to 190°C for a similar period would generate natural gas.

So much for a theoretical model. Let us now examine two basins where the geological histories have been different (Figure 33). In both the Paris and Los Angeles basins the *threshold*, or beginning of petroleum generation, has been dependent on depth and temperature, and duration of burial. In the Paris basin, the threshold was reached after 40 Ma when the early Jurassic rocks were buried to a depth of 1400 metres where the temperature was 60°C. The source rocks of the Los Angeles basin were buried to 2500 metres, where the temperature was 115°C, before the threshold was reached. It has been estimated that the petroleum in the Paris basin was generated over a period of 120 Ma; in contrast the generation of the petroleum in late Tertiary Los Angeles rocks, which were buried quickly to a much greater depth, took less than 10 Ma. Because the geothermal gradient, and the depth and duration of burial, can vary considerably from basin to basin, so can the time taken for petroleum to be generated.

6.4   Summary of Chapter 6

1   Petroleum occurs in liquid (crude oil), gaseous (natural gas) and solid (asphalt) states, and is composed of many hundreds of hydrocarbon compounds. There is conclusive evidence that petroleum is derived from organic matter which was buried within sediments in an oxygen-free environment to form source rocks.

2   The generation of petroleum begins with the conversion of the organic matter into kerogen during diagenesis. At higher temperatures and pressures, during categenesis, the larger hydrocarbon molecules are broken down into crude oil and natural gas.

3   The type of maturation product depends on the origin of the organic matter. Three kinds of kerogen have been identified. Type I is derived from marine algae, Type II from a mixture of marine organisms and Type III from terrestrial floras: Type II is the main source of crude oil.

4   Petroleum in generated in small amounts when temperatures are below 50°C with maximum production at 100°C; if heated to over 150°C even crude oil will break down to form natural gas. Such temperatures are acheived in the Earth's crust when source rocks are buried within sedimentary basins; the geological time that these rocks spend at critical temperatures will, in part, decide the volume of crude oil to be generated.

7 Migration and accumulation of petroleum

Study Comment. In this chapter we first consider how petroleum migrates through the strata following its generation and then we examine the types of structure in which petroleum deposits might be located. This study requires an understanding of the properties of sedimentary rocks and the factors controlling the movements of fluids through them together with geological structures like anticlines and unconformities.

7.1 Sedimentary basins

The formation of petroleum requires marine sediments that have been buried to considerable depths, so the obvious place to look for petroleum accumulations is in areas where thick successions of marine strata have accumulated in sedimentary basins. Some basins are to be found on the present-day continental shelves and along continental margins, while others are onshore basins that are the remains of old sedimentary basins. (An example of the latter is the Middle East basin; cf. Chapter 9.) Within such basins, petroleum can collect to form *oil fields* , which are areas where the rocks have become saturated with liquid hydrocarbons in commercial quantities. *Gas fields* are found where there are concentrations of gas in sedimentary rocks, and can be in areas where crude oil occurs and also where there are coalfields (Chapter 3.3).

7.2 Migration of petroleum

Although the generation of petroleum in source rocks is fairly well understood, its migration into other rocks is not. Two stages have been recognized in the migration process (Figure 34): *primary migration* is the movement of the hydrocarbons within and then out of the source rocks, and this is followed by *secondary migration* into and within the rocks where it accumulates, which are called *reservoir rocks*.

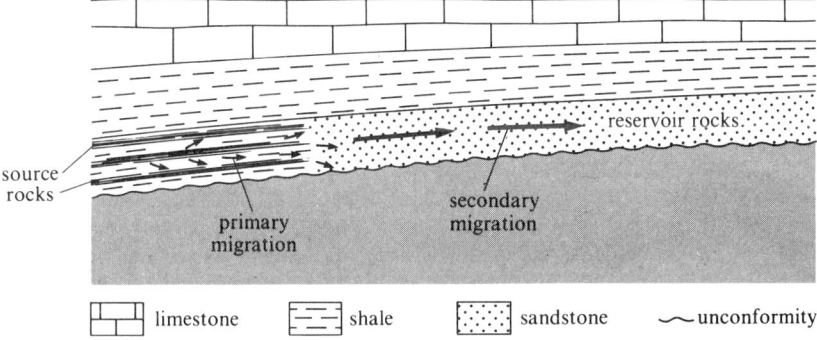

Figure 34 Primary migration of petroleum out of deeply buried source rocks that have reached maturation temperatures, followed by secondary migration into sandstone reservoir rocks.

The source rocks are mainly organic-rich shales (Chapter 6.2), which are very fine-grained and impermeable. So how can the generated petroleum move out of the source rocks? One current suggestion is that when kerogen is subjected to maturation temperatures, methane is formed which creates so much internal pressure within the source rocks that microfractures are formed which then permit the primary migration of the hydrocarbons. Having escaped out of the source rocks, the petroleum appears to move with greater freedom along joints, faults and bedding planes into a reservoir rock such as sandstone.

7.3 Reservoirs

The properties of a reservoir rock are very like those of an *aquifer* through which water can flow. It has porosity, and the petroleum is to be found in the *porespaces*, and also in fractures in the rock. it must also be *permeable* if the petroleum is to be extractable. Figure 35a shows that quite large pore spaces may exist between the rounded grains of quartz that form a sandstone. The more pore spaces relative to sedimentary particles there are in the rock, the greater the porosity of the rock. And if pore spaces join then the rock becomes permeable and fluids can move through it. During the evolution of a sedimentary basin, fluids such as water are constantly migrating through the pore spaces; migrating low-density crude oil can occupy these pore spaces, displacing most of the denser water, but with some water still left as a thin film around each grain (Figure 35b). Sandstones are common reservoir rocks for petroleum and hold 60 per cent of the world's resources.

Lithification can reduce the porosity and the permeability of sandstones, by the deposition of silica around quartz grains from migrating groundwaters (Figure 35c). However, sometimes migrating waters can increase porosity and permeability by dissolving limestones and widening fractures. Because of such fractures, limestones hold 40 per cent of the world's resources of petroleum.

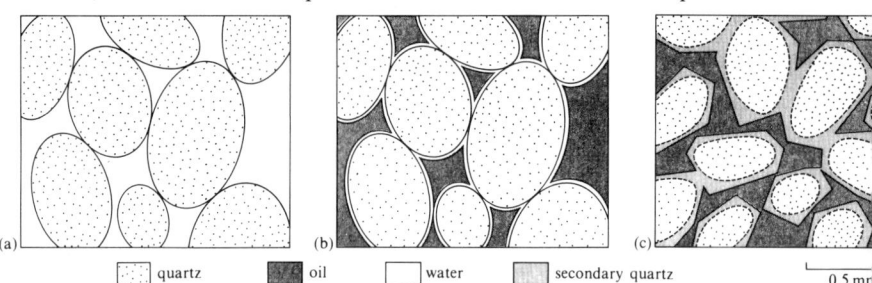

| quartz | oil | water | secondary quartz | 0.5 mm |

Figure 35 Idealized magnified thin sections through sandstones: (a) the pore spaces filled with water; (b) oil displacing the water during secondary migration, with water remaining in a thin film around the grains; (c) the reduction of pore spaces by deposition of secondary quartz around sand grains during lithification.

7.4 Traps

Petroleum migrates upwards from source rocks into a reservoir until a *seal*, such as an impermeable shale, stops it. In order for oil and gas fields to form, suitable geological structures must exist to trap the petroleum. Before we examine the different kinds of *trap*, consider the following five prerequisites for the accumulation of oil and gas to form a field:

1 Suitable source rocks must have been subjected to maturation processes;

2 Suitable reservoir rocks must exist;

3 Secondary migration of petroleum from source rocks into reservoir rocks must be possible;

4 The reservoir rocks must occur within large enough traps so that economic accumulations can form;

5 The escape of petroleum to the surface must be prevented by a suitable seal.

In some sedimentary basins, although source rocks have obviously produced petroleum, no suitable traps existed; in other areas, traps were formed long after the generation of petroleum, which had already escaped. So when carrying out an exploration survey of a basin it is necessary to work out the sequence of geological events, and once a trap is discovered it is essential to determine how and when it was formed.

There are three types of trap. *Structural traps* are formed as a result of the deformationof strata in the Earth's crust. *Stratigraphic traps* occur where there is a permeability barrier caused by lateral and/or vertical variation in sedimentary rock types. Some large petroleum traps have been formed by both structural and stratigraphic means and these are known as *combination traps*. Analysis has shown that 78 per cent of the world's crude oil resources are held in structural traps, 13 per cent in stratigraphic traps and 9 per cent in combination traps.

7.4.1 Structural traps

The *anticlinal trap* is the commonest form of petroleum accumulation, and these traps acount for the largest volume of the world's resources. Anticlinal traps are common in regions where the crust has been subjected to horizontal compression and they can be large structures. Figure 36 shows an anticline; in three dimensions the structure is a *dome*, in which the rocks dip downwards from the centre in all directions.

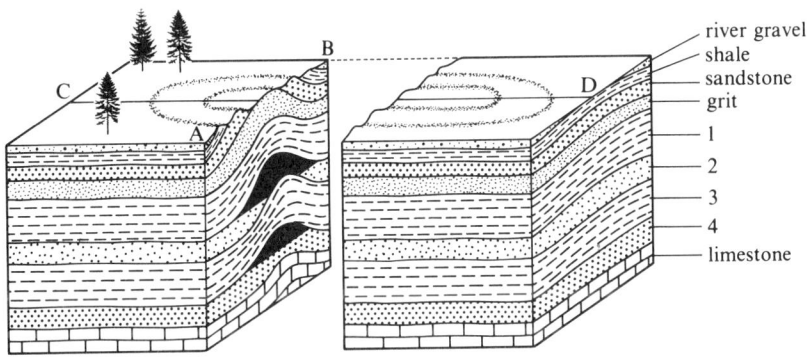

Figure 36 An idealized block diagram showing a dome structure in folded sedimentary strata which form concentric ridges at the surface. Section A–B shows an anticlinal structure with oil trapped at two levels.

Two separate traps are shown in Figure 36, and they are there because of the vertical continuity of the dome: in California one such structure had 25 separate sandstone reservoirs stacked vertically one above the other.

The limiting factors that determine the volume of petroleum in an anticlinal trap include the following:

1 The size of the structure: anticlines can be hundreds of kilometres long and thousands of metres in height;

2   The volume of the source rocks that have undergone maturation after the structure was formed;

3   The porosity of the reservoir rocks;

4   The thickness of the reservoir(s);

5   The amount of *closure* on the structure, that is the position of the *spill-point* relative to the height of the arched reservoir that lies above this level.

The last point is illustrated in Figure 37. Petroleum migrates up-dip from the source rocks because of its low density and gradually fills a trap. Any gas present will always lie above the oil because it has a lower density. A stage is reached when oil can escape from the trap at the spill-point; such a trap is said to have only a limited closure.

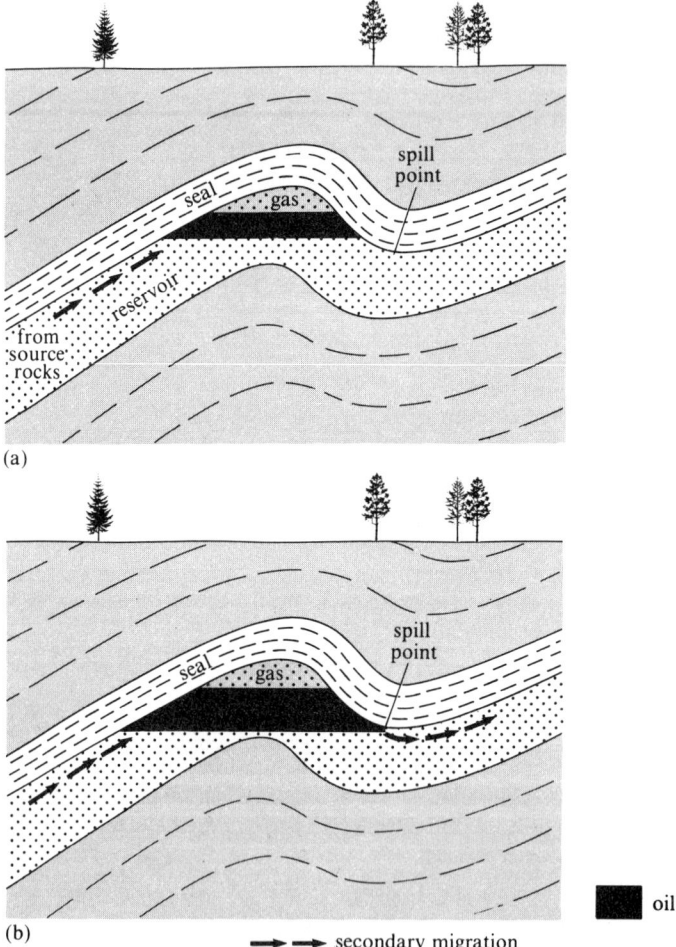

Figure 37   The migration of oil and gas into and out of an anticlinal trap. (a) During the early phase of secondary migration the lower level of the oil in the reservoir has not reached the spill-point. (b) Later, the spill-point has been reached and oil can flow further along the reservoir.

Anticlinal structures can also be formed in non-compressional areas of crust and may result from a *drape* of sedimentary rocks over an uplifted or faulted older rock; an example of this structure is provided by the Forties field in the North Sea. Other anticlinal structures can be formed by *salt plugs*. These develop in regions underlain by evaporite beds rich in *halite* (salt, NaCl). When it is deeply buried and under high pressure, salt behaves as a plastic material within denser rocks. Its relatively low density means that it tends to rise towards the surface in elongated cylindrical plugs or pillars of salt. Salt plugs, which are commonly about 1 km in diameter, are impermeable to petroleum, and in suitable areas the upward moving plug will pierce the overlying strata, including reservoirs, with the result that traps are formed (Figure 38). Because some salt pillars can be up to 10000 metres in height there may be many reservoirs and hence many traps. Associated with the upward movement of salt are faults in the surrounding rock, and traps can also be formed by these faults (see below). Above the salt plug is the *cap rock*, a series of layers of *anhydrite, gypsum* and limestone, which can contain cavities and show evidence of fracturing. In some instances the cap rock acts as a reservoir for oil and gas. The exact origin of the cap rock is unknown, although there are several hypotheses for its formation.

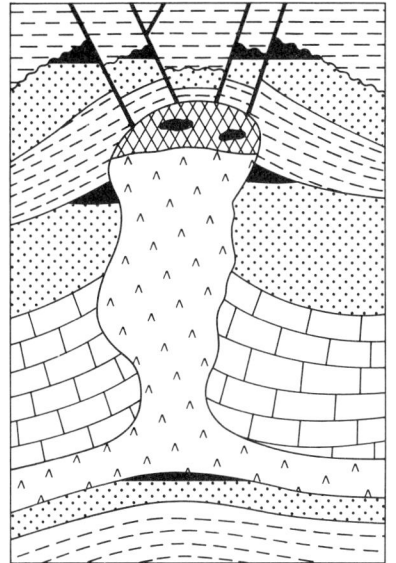

Figure 38 A salt plug and the associated oil traps.

cap-rock

shale

limestone

sandstone

salt

oil

~~~ unconformity

— faults

Fault traps are formed when an inclined bed of reservoir rock is brought into contact with impermeable strata by a fault. Some fault traps are illustrated in Figure 39, where the reservoir rocks are beds of sandstone, above and below which are thicker beds of impermeable shale that act as seals. Where the complete thickness of the sandstone reservoir has been faulted against a shale there is said to be unlimited closure; in other cases the closure is limited and there are spill-points.

So far we have discussed traps that have been formed mainly as a result of folding or faulting of rocks. The next category of traps normally occurs in relatively undeformed sedimentary basins.

7.4.2 Stratigraphic traps

We briefly considered deltaic deposits in Chapter 3, and it was pointed out that

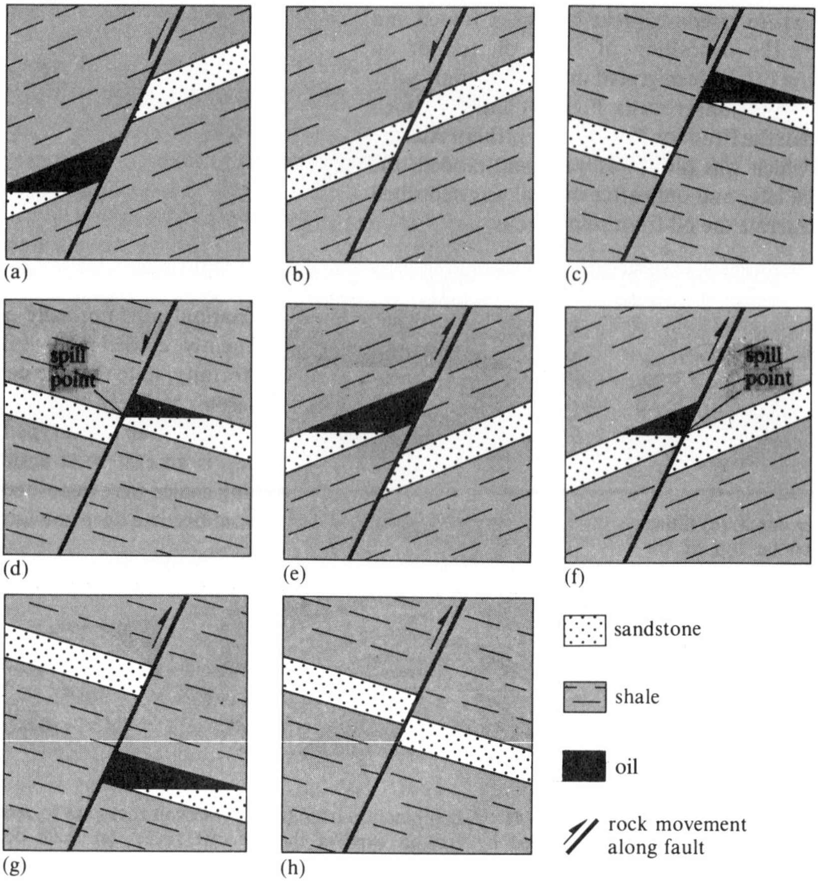

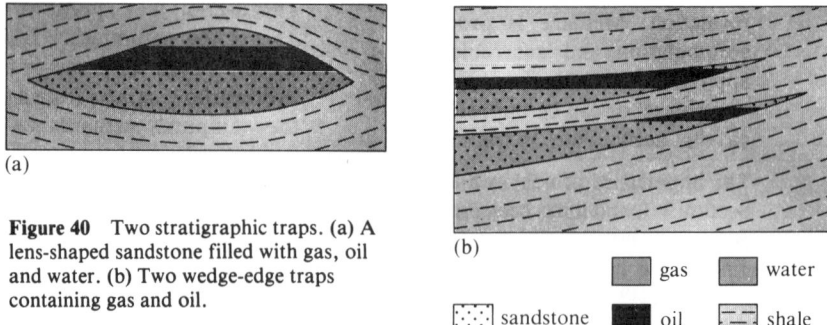

Figure 39 Various kinds of fault trap. You will see that petroleum is trapped in all the structures except (b) and (h) in which movement along the fault has been insufficient to seal the sandstone against the shale and consequently petroleum can escape up-dip. There is limited closure on traps (d) and (f) and unlimited closure on (a), (c), (e) and (g).

Figure 40 Two stratigraphic traps. (a) A lens-shaped sandstone filled with gas, oil and water. (b) Two wedge-edge traps containing gas and oil.

sediments vary laterally in composition. From modern analogues we know that a sandy bed deposited near shore can pass laterally into muds deposited in deeper water. Similarly some sandstones are discontinuous within shale deposits and appear like lenses or wedges in cross-section (Figure 40a; note that gas, oil and water are shown in separate layers and this is always the relationship because of their relative densities). When wedge-shaped reservoirs are inclined, petroleum can be found at the upper end in what are known as *wedge-edge traps* (Figure 40b). Reefs similar to modern coral reefs have existed in the geological past. Like some modern reefs the fossil reefs can be circular in outline and are composed of limestone made up of corals and shell fragments, which are later dissolved to give *secondary porosity*. When reefs become buried underneath impermeable sediment, such as mud, which later becomes compacted to form shale, they form *reef traps*. Such traps are restricted to rocks formed in former tropical and subtropical climates and acount for only 3 per cent of the world resources of petroleum.

The final example we shall examine is the *unconformity trap*. A period of uplift and erosion results in tilted beds below the unconformity overlain by nearly horizontal beds, which can act as a seal. Major unconformities extend over large areas of the Earth and thus might be expected to have trapped large accumulations of petroleum. So what is the main reason why unconformity traps contain only 4 per cent of the world resources? Large volumes of petroleum probably escaped when the sea floor was uplifted, so although the younger strata above the unconformities form suitable seals, there is no oil left to seal in.

7.4.3 Combination traps

Combination traps normally occur where reservoir rocks have been folded and subsequently eroded and sealed by the deposition of shales on top of an unconformity; note that the petroleum can have migrated into the trap only after the reservoir had been sealed by the much younger rocks. Figure 41 illustrates such a combination trap of the type found in the Prudhoe Bay field in northern Alaska. The trap is an east–west anticline which was tilted westwards and eroded; the overlying shales were deposited much later and only after this seal was deposited did burial become deep enough to generate the oil from source rocks.

7.5 Natural gas

This section may appear to be a digression at this stage, but it is important to discuss how natural gas is related to other hydrocarbons. It is possible to trace a

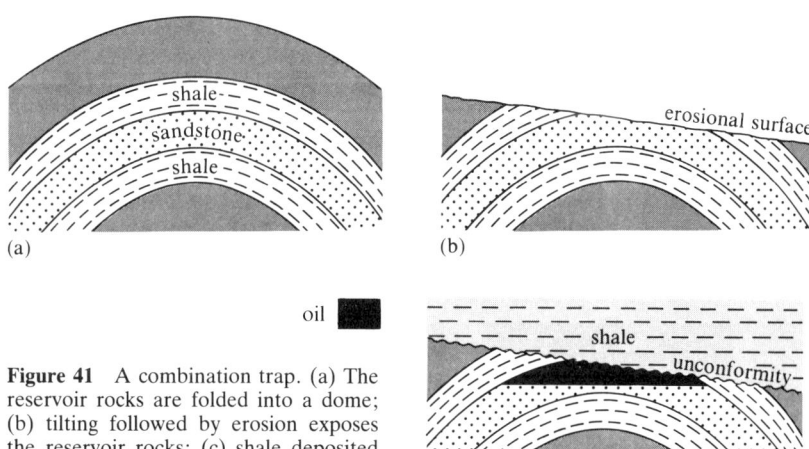

Figure 41 A combination trap. (a) The reservoir rocks are folded into a dome; (b) tilting followed by erosion exposes the reservoir rocks; (c) shale deposited on top of the unconformity acts as a seal and creates a trap.

crude oil back to its source by comparing its geochemical analysis with that of hydrocarbon extracts from a particular source rock. But what of a gas source? If the gas is purely methane then it is virtually impossible to determine its exact source and hence its origin.

There are different kinds of natural gas, with differnet compositions and modes of formation. Methane is produced by bacteria in the very early stages of accumulation and anaerobic decay; it occurs at low pressures which are approximately atmospheric pressures and normally escapes into the atmosphere as 'marsh gas' or can be trapped at shallow depths. Now let us consider the natural gas that accumulates in traps, with or without crude oil.

The importance of coal as a source of natural gas has already been discussed in Chapter 3.3, and it has been established that coals are able to generate large enough volumes of methane for commercial accumulations to form. Although small amounts of liquid hydrocarbons are generated during coal formation, very few commercial oil accumulations have been traced to a coal origin. It has been suggested that if liquid hydrocarbons are formed during the maturation of coal, only limited primary migration is possible because oil is rapidly absorbed by coal.

Because coal is formed from terrestrial organic matter it has a strong similarity to Type III kerogen (Table 11), which yields gas rather than oil during maturation. If you refer back to Figures 29 and 30 you can see that during the maturation of any kerogen, natural gas and crude oil can form simultaneously. Natural gas dissolves in crude oil under pressure, but comes out of solution as soon as the pressure is released (e.g. by a borehole). In the later stages of maturation the crude oil can be converted into wet gas, which is a mixture of oil and gas, and finally with increased temperature all the oil is converted into dry gas.

Like crude oil and coal, natural gas has generally been regarded as having had a biological origin, but some recent evidence indicates that methane is present in the Earth's mantle. It is suggested that vast amounts of methane lie deep in the Earth, and as petroleum reserves begin to decline during the next few decades exploration may well be directed towards testing this theory.

7.6 Solid petroleum and oil shale

Crude oil that has migrated through a reservoir can sometimes reach the surface, where the volitiles escape and the rock becomes impregnated with a hydrocarbon residue; such a rock is known as an *oil sand* or *tar sand*. The Athabasca oil sands of Alberta, Canada, are such a deposit, and they contain 600×10^9 barrels of asphaltic hydrocarbons. In this case the oil first accumulated in Devonian reef traps from which it escaped and entered a permeable Cretaceous sandstone, with a westerly dip, which was later exposed at the surface. Because of this exposure, the oil lost most of its volatiles and became too viscous to flow; the rock has to be heated to extract the oil. The Athabasca oil sands are mined by open-cast methods and then treated with steam and hot water to separate oil from sand and pebbles; the cleaned oil is then sent to refineries.

Pitch lakes are large deposits of asphalt formed from oil that has seeped to the surface, accumulated in large depressions and become solid. Examples include deposits in Trinidad, and the Bermudez Lake in Eastern Venezuela with an estimated reserve of 6.1 million tonnes.

A shale that contains 2.5 per cent or more of unaltered kerogen is termed an *oil shale*, and when heated to about 500°C it yields crude oil; because it has to be produced by heating, the product is termed *'synthetic' crude oil*. The energy required to heat the oil shale to that temperature is about 1 kJ per gram of rock and the calorific value of kerogen is about 40kJg^{-1}; therefore if the kerogen forms only 2.5 per cent of the rock its total calorific value is taken up heating the shale. To be economic the lower limit is frequently taken as 5 per cent kerogen in the USA, and this yields 25 litres of synthetic crude per tonne of rock.

There are major economic problems in the extraction of synthetic crude oil, but it is still being distilled in the Baltic region of the USSR and by the Chinese in Manchuria. The Lower Carboniferous oil shales of the Midland Valley in Scotland reached a peak production of 3.35 million tonnes in 1913, with a distillation rate of about 100 litres per tonne of shale. Oil shales have been deposited in lakes, shallow seas and lagoons from Precambrian to Cenozoic times, and some of the largest deposits are located in the American states of Utah, Wyoming and Colorado. It has been estimated that the oil shales could yield about 18000×10^9 barrels of oil, but because of the escalating costs, and a shortage of available water for processing, most of the projects involved with the production of synthetic crude oil were temporarily abandoned in the early 1980s.

7.7 Summary of Chapter 7

1   Crude oil and natural gas can accumulate in commercial quantities only when the correct sequence of prerequisites has occurred. Suitable reservoir rocks must exist within a trap that is sealed, and the source rocks ust then be subjected to maturation processes; the resultant petroleum then migrates into the trap as either oil or gas, or both.

2   Primary migration describes the movement of hydrocarbons within the source rocks and secondary migration is its subsequent movement into reservoir rocks. The reservoir rocks are normally permeable, fractured sandstones or limestones.

3   Geological structures that are sealed against the escape of petroleum by rocks such as shale and salt deposits are known as traps. Structural traps are formed by the deformation of sedimentary rocks and include domes which are anticlinal in vertical section, and fault traps which are formed when inclined reservoirs are moved by faulting against impermeable beds. Other examples include traps associated with salt plugs.

4   Stratigraphic traps are formed by lateral changes in the permeability of reservoir rocks and include wedge-edge, reef and unconformity traps. Combination traps are formed when folded or faulted reservoirs are overlain unconformably by shales, which act as a seal.

5   Natural gas can be associated with crude oil as wet gas or can exist on its own as dry gas; dry methane gas can also be derived from coal.

6   Solid petroleum exists in the form of oil sands and tar sands, pitch lakes and oil shales; it is possible to derive synthetic crude oil from oil shales by heating, but the cost of the energy input is at present too high to encourage commercial exploitation.

8   Petroleum exploration, evaluation and recovery techniques

> **Study Comment.** In this Chapter we consider some of the methods used to detect possible traps, find out if they contain petroleum and estimate the volume. The evaluation of petroleum resources depends in part on the use of structure contours and isopachyte maps. Then we examine different methods of estimating the world's reserves and resources of oil and gas and consider methods used to improve the recovery of oil from traps.

8.1   Exploration techniques

Surface geological mapping of rock outcrops, for example the surface expression of the dome in Figure 36, will provide some data on the possible subsurface structure but, in order to establish if such structures are continuous to depth, the subsurface geology has to be identified by geophysical exploration methods. Seismic reflection profiling (Chapter 4) is the main method used in exploration for oil traps. This is because seismic techniques have high accuracy and resolution and great depth of penetration, and provide data that are relatively easy to translate into geological terms.

As you should recall from Chapter 4, the seismic reflection technique requires that a source of sound energy is produced and that such energy is efficiently directed downwards into rocks of the Earth's crust, where it is reflected at interfaces between rocks with different acoustic properties. The reflected energy is then received at a series of detectors placed at or near the surface of the land or towed through the water at sea. The time taken for a seismic impulse to pass from the energy source to the detectors via a series of reflecting interfaces can be used to construct a picture of the geological structures at depth.

A seismic survey in 1965 in the central North Sea revealed a broad anticlinal structure, but it was not possible to determine the ages of rocks or confirm the existence of petroleum in such a seismic profile without exploratory drilling. In this case drilling in the early 1970s revealed an oil trap in the 55 Ma old early Tertiary sandstones over 2 km below the sea bed. This is the Forties oil field (cf. Chapter 9).

8.2   Trap structure and the recovery of oil

When an oil field is discovered below the surface, it is important to map out the structure in order to ascertain its three-dimensional shape; from these data it is then possible to determine accurately what type of oil trap it is. As an example of this work we will consider the Leduc field of Alberta, Canada. In order to determine the shape of the upper surface of the limestone reservoir, *structure contours* had to be drawn from data obtained from wells drilled into the structure.

Figure 42 provides a structure contour plan for the top of the limestone together with the borehole data. You will notice that the contours have revealed a domal structure, elongated in a north-west to south-east direction. The next question is whether the structure is anticlinal in cross-section or not?

What information is required to find out whether this is true for the limestone reservoir?

We need to know the thickness of the limestone in each well. These data are shown in Figure 43 together with an isopachyte map (i.e. lines of equal thickness) of the limestone.

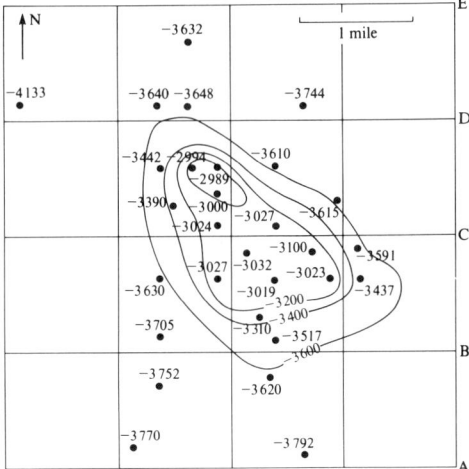

Figure 42 Structure contours and borehole data for the depth (in feet) below sea-level to the top of the oil-bearing limestone in the Leduc field.

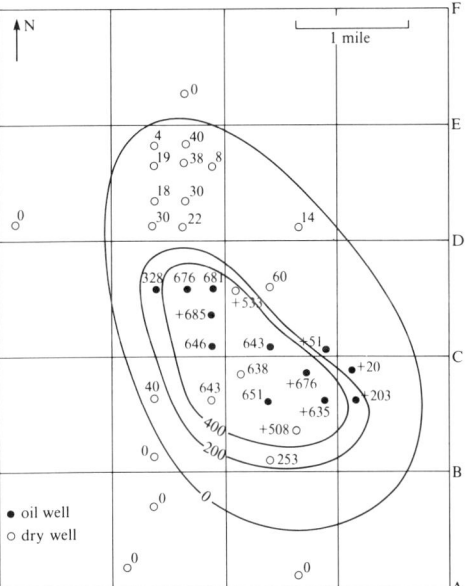

Figure 43 Borehole data and isopachytes for the thickness (in feet) of limestone drilled in each well. Figures such as +203 are minimum values and mean that the wells did not reach the base of the limestone.

The isopachytes show that the limestone is thickest at the centre (600 feet or more), thinning to zero at the edges. They have a similar shape to the structure contours for the top of the mound, i.e. roughly circular. This suggests a mound of limestone with a convex upper surface and the structure of a reef. The oil trap is therefore stratigraphic in nature with the oil wells clustered towards the thickest part of the reef and aligned along the NW–SE axis of the structure.

Having described the structure of the oilfield, the next step is to calculate the volume of oil in the field.

The formula N=VRP is used to calculate the number (N) of barrels of oil (1 barrel is $0.159m^3$) recoverable from the reef, given that the porosity (P) of the reservoir is 8 per cent and the estimated recovery (R) is 40 per cent. This calculation gives the volume of recoverable oil as 19×10^7 barrels.

We will be dealing with the North Sea oil fields in Chapter 9, but in order to place the reserves of the Leduc field in perspective, let us have a quick forward look at the estimates for two of the North Sea fields. The same method was used to estimate that the Brent field, discovered in 1971, had 10^9 barrels and the Forties field, discovered a year later, had 2.4×10^9 barrels of recoverable reserves. So the Leduc field, with only 19×10^7 barrels, is relatively small.

8.3 Primary, secondary and enhanced recovery methods

Oil in a reservoir resembles water in a *confined aquifer* and is at pressure that can be defined by the *piezometric surface*. Once the first exploratory well has penetrated a penetrated a reservoir, the fluids begin to move and if the pressure is great enough the petroleum will rise up the borehole and reach the surface. As the pressure is released, any gas dissolved in the oil will come out of solution and rise and escape. Then the pressure of the petroleum remaining in the reservoir begins to fall. It has been calculated that during this *primary recovery* only 30–40 per cent of the petroleum in the reservoir will be brought to the surface, because the fall in pressure and the loss of dissolved gas cause increases in both the surface tension and the viscosity of the oil, i.e. it becomes 'thicker' and will not flow so readily. Since the 1940s petroleum engineers have developed *secondary recovery* techniques which can increase recovery to about 50 per cent. The techniques include the injection of natural gas into the reservoir *above* the oil forcing the oil downwards, and flooding with water *below* the oil, forcing it upwards.

In order to improve recovery still further, new techniques have been developed and these have been termed *enhanced recovery* methods. Such techniques include the addition of chemicals to the injected water and the injection of steam in order to reduce the viscosity of the crude oil. In spite of these enhanced recovery techniques it has been estimated that 50 per cent of the oil in the reservoirs of the North Sea will have to be left there. So further research is under way, aimed at the recovery of some of this 300×10^9 US dollars worth of oil (1983 prices).

Some exhausted wells in the USA have been injected with bacteria in molasses and sealed off. This results in an increase in the flow of petroleum, owing to the pressure of the gas produced by the bacteria. Present research includes the genetic engineering of bacteria that thrive on crude oil. It is predicted that the injection of suitable bacterial strains into the reservoir will produce gases such as hydrogen and methane to increase the pressure; other strains can produce chemicals, such as alcohols, that will act as solvents, and yet others can produce chemicals that will reduce the viscosity. There is considerable optimism in the early 1980s that these new enhanced recovery methods will boost the volume of recoverable oil and therefore prolong the 'life' of oil fields. But an equally important factor that will determine how long the recoverable gas and oil will last is the rate of consumption, which we will discuss in Chapter 9.4.

8.4 Summary of Chapter 8

1   The main methods of determining whether an area has traps that may contain petroleum are geological mapping and seismic reflection profiling. Once detected, a potential trap can be mapped in detail by drilling, and the well data are then used to map structure contours and isopachytes. Once the volume of the reservoir rock, its porosity and estimated percentage recovery has been determined, it is possible to estimate the volume of oil that can be recovered.

2   Primary recovery methods produce only 30–40 per cent of the reserve; this can be boosted to 40–50 per cent by secondary recovery techniques of pumping water and gas into the reservoir. Enhanced recovery is intended to increase recovery still further, mainly by the injection of chemicals into the reservoir to reduce viscosity; another exciting prospect is the use of bacteria to produce gas or solvents within the reservoir. Any improvement in the percentage recovery will have important bearing on the estimates of total recoverable reserves.

9 Petroleum in Britain, and its world setting

Study Comment. Chapter 9 looks first at the locations of the gas fields and oil fields on the British mainland, and then examines how the discovery of an onshore gas field in the Netherlands led to the exploration and development of the North Sea resources. After a detailed look at the geology of the North Sea basin, we consider how the Middle East basin exemplifies the ideal conditions for the generation and trapping of very large volumes of petroleum. Finally we look at the effect of the rate of consumption on petroleum reserves and at new techniques that are being developed to extend the life of petroleum products by various conversion processes based upon coal.

9.1 Onshore explortion in Britain

Oil was first discovered in commercial amounts onshore in Britain in 1939, when the D'Arcy Exploration Company, now British Petroleum, found the Eakring oil field 25 km north-east of Nottingham in the Coal Measures concealed below Permo-Triassic rocks (Figure 44). During onshore exploration between 1939 and 1980 more than 900 wells were drilled, which resulted in the discovery of 26 small oil fields. Up to 1980 these had yielded nearly 25×10^6 barrels of crude oil: to put this into perspective, Britain consumes about 610×10^6 barrels per year (1983), so the *total* production so far would last only 15 days.

Perhaps the most exciting prospects in Britain at present are in southern England. The Wytch Farm oil field, near Corfe Castle in Dorset, was discovered in December 1973 and production is from two porous Triassic and Jurassic sandstones. This field is believed to be similar in size to the Argyll oil field in the North Sea, and by the end of 1982 had produced nearly 4 million barrels of oil.

9.2 Onshore to offshore

Geologists who examined rocks of onshore oil fields of the Gulf of Mexico and Caspian Sea areas were able to predict that the oil fields extended seawards. Offshore drilling in the shallow waters in the early part of the century proved that, but in spite of this knowledge the intensive offshore drilling of sedimentary basins only started inthe 1950s. The main reasons for this delay were two-fold: (i) the vast onshore oil fields had been sufficient to meet demand and (ii) the move to offshore drilling involved a new and hostile environment with problems of weather, sea conditions, sea-floor stability and greatly increased costs. All of this lead to major new developments in exploration and engineering technology which are continuing today as exploration extends into even deeper waters.

9.2.1 North Sea petroleum

In 1959 a well near Groningen in the Netherlands was drilled to test a domal structure, and gas was found in Lower (basal) Permian rocks at around 3000 metres (Figure 45). The reservoir ranges in thickness from 100 metres in the south to 200 metres further north, and consists of coarse reddish river-lain gravels and brown wind-blown sandstones with porosities ranging from 12 to 20

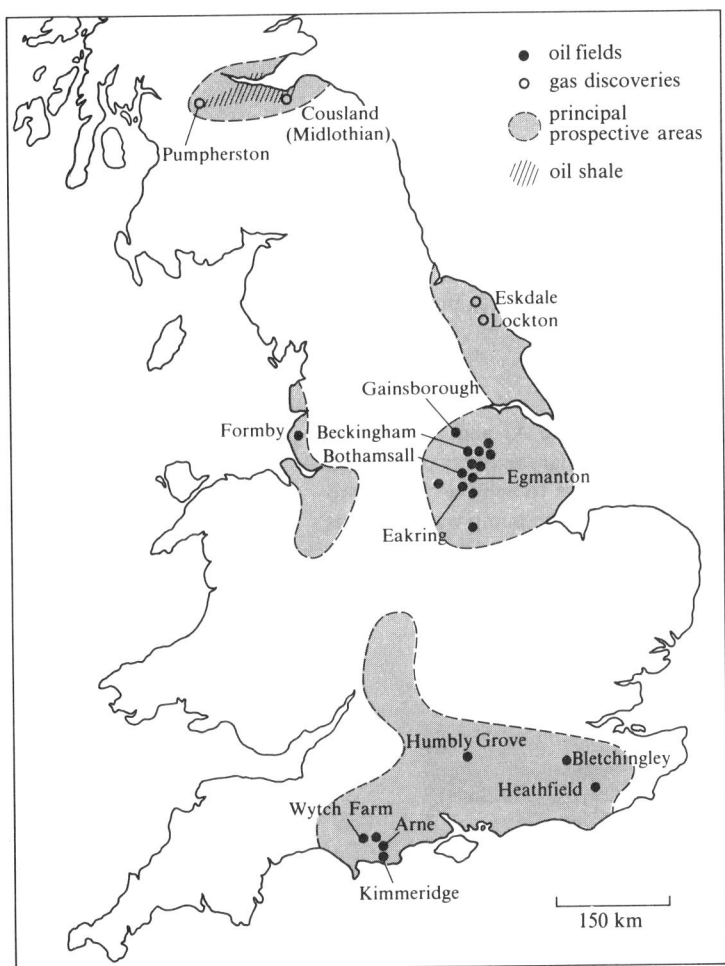

Figure 44   The locations of the onshore oil and gas fields in Britain.

per cent; it is overlain by an evaporite seal. This field, with an estimated reserve of 1500×19^9 m^3, is one of the largest known gas accumulations in the world.

The Groningen gas field made geologists reconsider the significance of the North Sea basin because, as you can see on a geological map, in north-east England there is a similar Permian—Carboniferous succession of rocks, including the Permian sandstone that forms the reservoir in the Groningen gas field. The problem was this: did these rocks extend outwards beneath the North Sea, and was the Permian evaporite there to form the seal? There was only one way to find out: offshore drilling. Geophysical surveys started in the North Sea as early as 1962, and it was logical to look for structures similar to that found at Groningen. The first discovery to be made was the West Sole gas field at a water depth of 27–30 metres, with the 150 metres thick Permian sandstone reservoir at

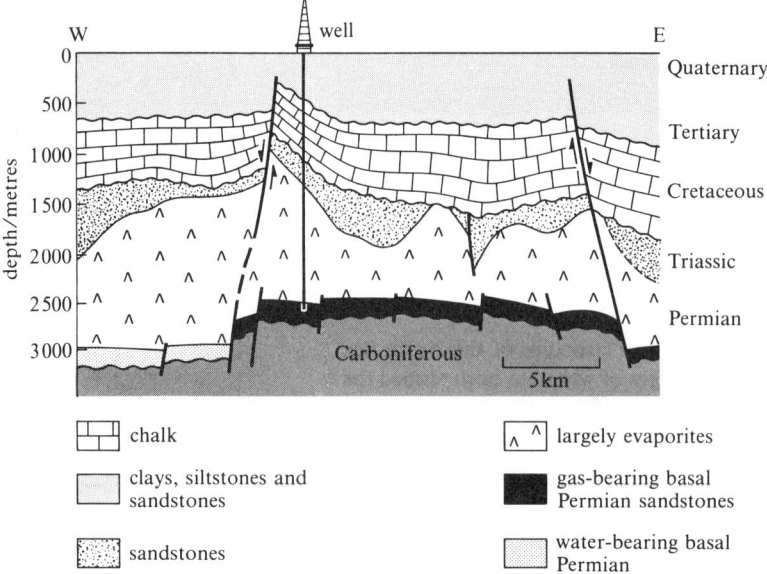

Figure 45 A section through the Groningen gas field; the reservoir in the basal Permian sandstone is anticlinal and faulted.

a depth of about 3000 metres. The structure turned out to be an elongated and faulted dome 19 km long and 5 km wide trending north-west to south-east within a stratigraphic succession of Quarternary, Jurassic, Triassic, Permian and Carboniferous rocks. The gas is almost pure methane and was probably derived from the Coal Measures beneath. The similarities between the Groningen field and this offshore gas field in the southern North Sea illustrate the point that once a good reservoir rock is discovered, it becomes a prime target for drilling in the search for other fields.

9.2.2 Structure of the North Sea basin

Beneath some shallow continental shelf seas are areas that have been subjected to long periods of subsidence and sedimentation. The North Sea is such an example, and sedimentation has taken place there for the last 400 Ma since the beginning of the Devonian. The key to our understanding of the structure of the North Sea came with the concept of plate tectonics. The stages of what has been termed the 'rift–drift' sequence in the formation of a new ocean are illustrated in Figure 46. Stages 1 and 2 illustrate the rifting phases, when tensional forces cause normal faulting of a continental plate to give elongated depressions, which are invaded and filled by seawater. Stages 3 and 4 show the formation of a new ocean, with the onset of seafloor spreading and drifting. We have examples of these stages in the East African rift valley (stage 1), in the Gulf of Suez (stage 2), the Red Sea (stage 3) and the Atlantic Ocean (stage 4).

In some cases the 'rift-drift' sequence stopped before it reached the stage that resulted in the formation of a new ocean underlain by oceanic basaltic crust. In the North Sea the sequence stopped at stage 2. At this time, during the Permian, the area was accumulating river- and wind-deposited sandstone reservoir rocks in the elongated central troughs, which are known as *grabens*. Subsequently the basin accumulated evaporite deposits in very shallow seas during the Permian,

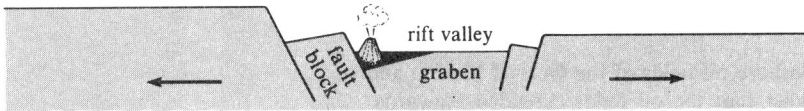

stage 1: rifting

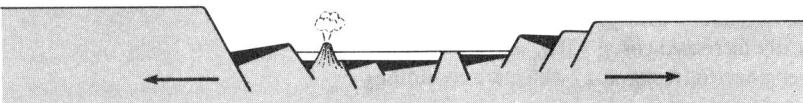

stage 2: further rifting; marine invasion

stage 3: new ocean

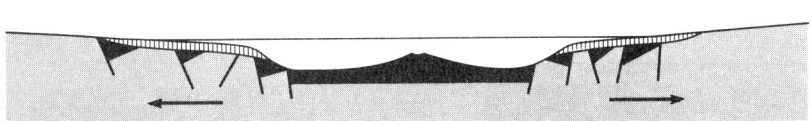

stage 4: enlarged ocean

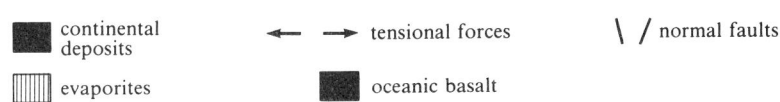

Figure 46   The rift—drift sequence, which leads to the formation of an ocean; in the North Sea the process stopped at stage 2.

and there followed normal marine sedimentation during the Mesozoic–Cenozoic which filled the original rift structure. This filled-in structure is illustrated in two cross-sections through the North Sea basin in Figure 47 (p. 81).

Geophysical exploration and drilling for oil and gas have revealed the complicated geological history of the North Sea basin, and Figure 48 shows several cross-sections of oil and gas fields. These illustrate the main types of petroleum trap in the North Sea basin while Table 11 analyses the geological details of these fields. The unconformity between the Devonian and Permian in the northern North Sea basin is important because the Carboniferous Coal Measures are not present in this area. Therefore there appear to be no source rocks for gas below the Permian. It can be concluded that the distribution of Coal Measures, which are restricted to the southern North Sea basin, explains the distribution of dry gas in the North Sea. The oil and gas of the northern North Sea basin are derived from the maturation of kerogen in marine shales.

In the North Sea basin there is a variety of reservoir rock types; these include Permian sandstones (West Sole and Indefatigable), Permian *dolomites* and Permian sandstones (Argyll), Jurassic deltaic sandstones (Piper and Brent), Cretaceous and Tertiary chalky limestones (Ekofisk), and Tertiary deep-water sandstones formed in *submarine fans* (Forties). Sands originally deposited on the continental shelf are sometimes transported and redeposited in deeper water where they form submarine fans, which can later be sealed by the deposition of normal deep-water shales. All these reservoir rocks were identified after drilling, but what of source rocks? The main source rocks of the crude oil in the northern North Sea are thought to be Jurassic Kimmeridge shales which also outcrop on the southern coast of England. In order to place the British onshore and offshore deposits in a global perspective, we will now examine the world's most prolific petroleum area. Table 12 (p. 84) summarises the estimates of recoverable oil and gas reserves in the U.K. continental shelf.

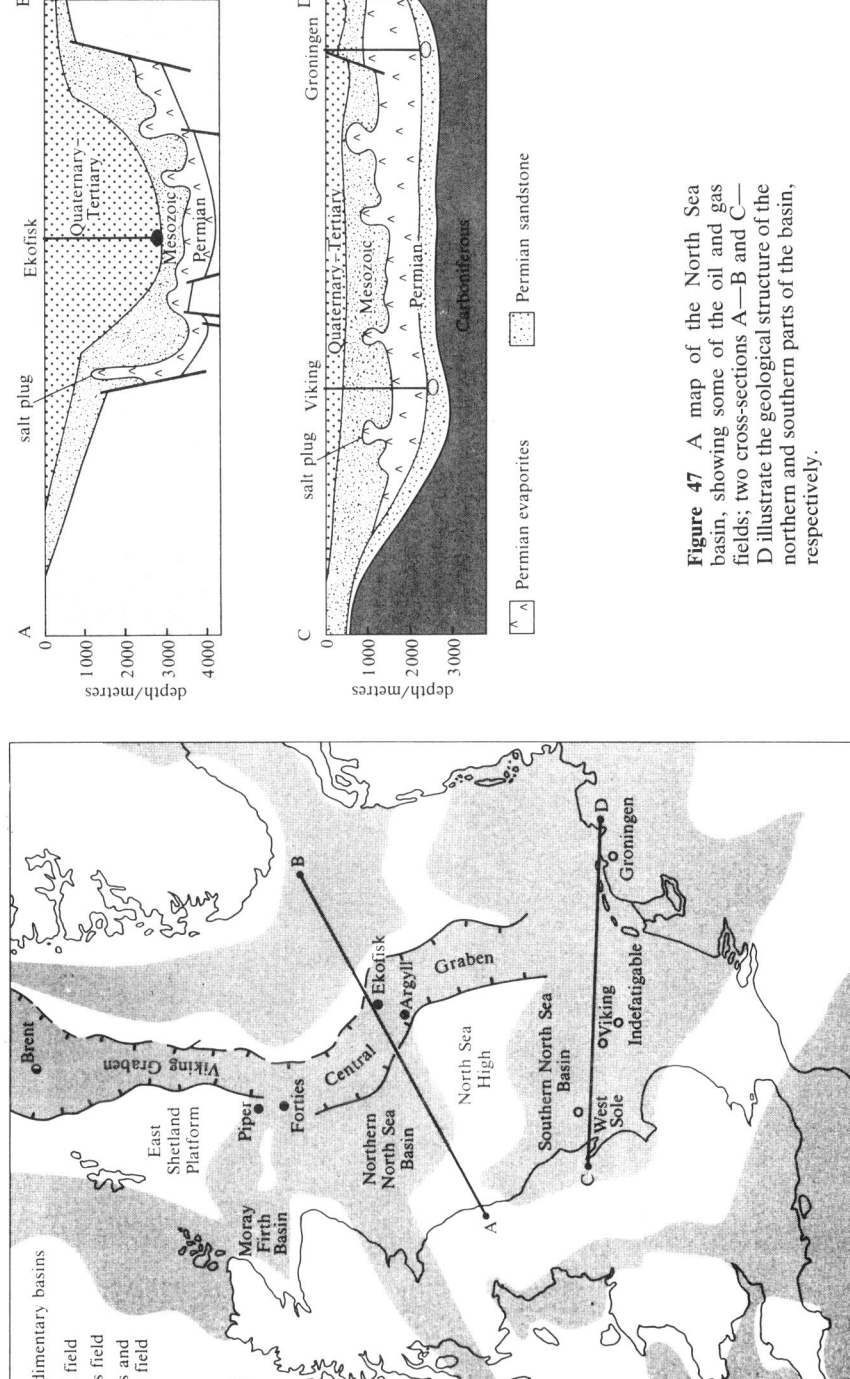

Figure 47 A map of the North Sea basin, showing some of the oil and gas fields; two cross-sections A—B and C—D illustrate the geological structure of the northern and southern parts of the basin, respectively.

82

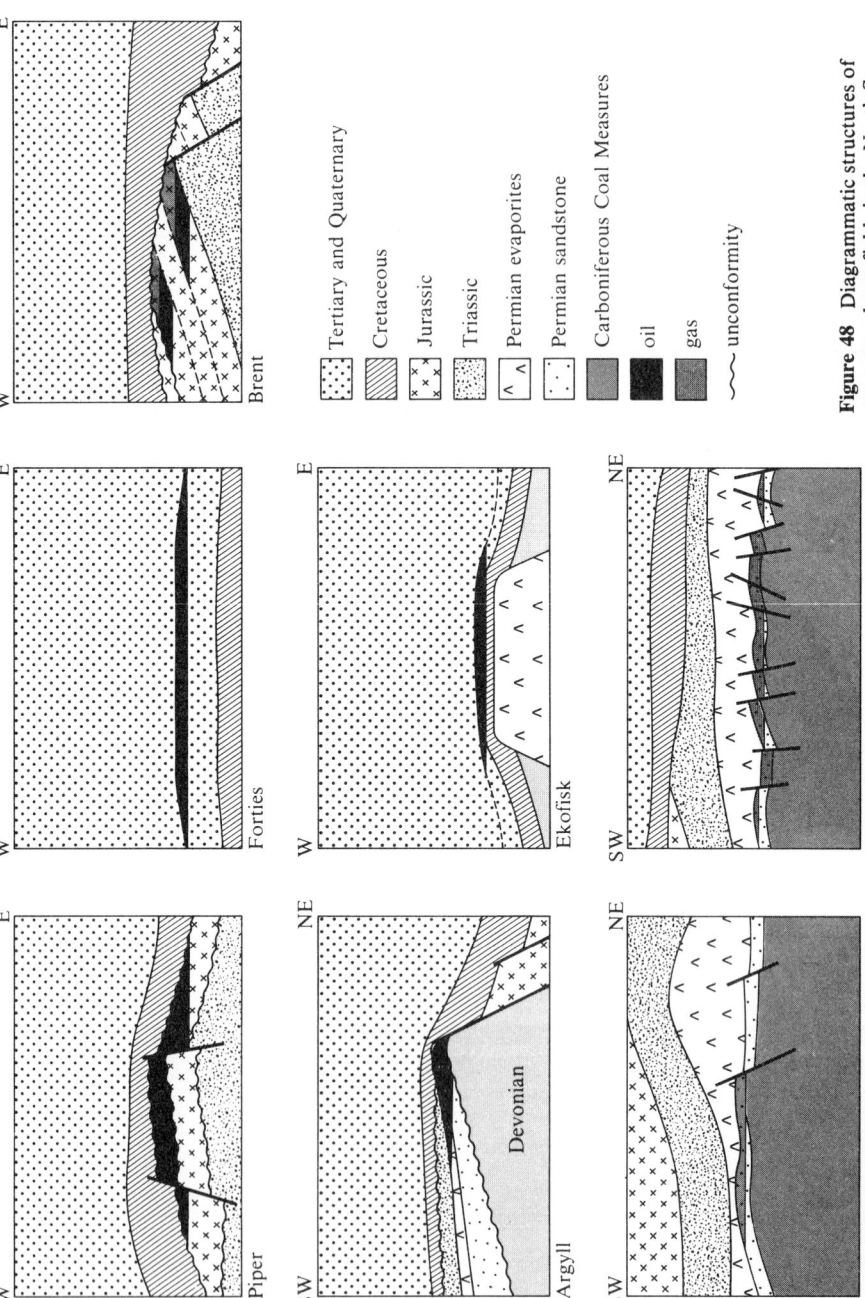

Figure 48 Diagrammatic structures of seven petroleum fields in the North Sea

	West Sole	Piper	Ekofisk	Indefatigable	Argyll	Brent	Forties
Location in North Sea	southern North Sea basin	east of Moray Firth basin	Central Graben	southern North Sea basin	Central Graben	Viking Graben	between Viking and Central Graben
Structure of field	faulted low dome	three tilted and folded fault blocks below unconformity	low dome above evaporites (salt plug)	faulted low dome	tilted fault block with unconformities	tilted fault block below unconformity	low dome (drape)
Type of reservoir	sandstone	deltaic sandstones	chalky limestone	sandstone	dolomites and sandstone	deltaic sandstones	deep-water sandstones (submarine fans)
Type of reserve	gas	oil	oil	gas	oil	oil and gas	oil
Age of reservoir	Permian	Jurassic	Cretaceous–Tertiary	Permian	Permian	Jurassic	Tertiary

Table 11 Details of the North Sea oil and gas fields illustrated in Figure 48.

Table 12 Estimates of Recoverable Oil and Gas Reserves in U.K.C.S. at end November 1984.

Each estimate is classified under two headings:
  Fields in production or under development
  Other significant finds not yet fully appraised

Category	Proven[*]	Probable[*]	Proven plus Probable[*]	Possible[*]
Oil (10⁶ tonnes)				
$Oil\ (10^6\ tonnes)$				
A.	1450	225	1675	250
B.	50	275	325	400
Total Initial Reserves	1500	500	2000	650
Less cumulative production to 1984	698			
TOTAL REMAINING RESERVES	800	500	1300	650
$NATURAL\ GAS\ (10^9 m^3)$				
1. *Gas from dry fields*				
A.	901	62	963	51
B.	156	210	366	218
2. *Gas from condensate fields*				
A.	40	8	48	14
B.	–	275	275	320
3. *Gas from oil fields*				
A.	130	17	147	11
B.	3	28	31	28
Total initial reserves	1230	600	1830	642
Less cumulative production	504			
TOTAL REMAINING RESERVES	725	600	1325	642

Notes: [*] The terms proven, probable, and possible are given the internationally accepted definitions.

> PROVEN – THOSE RESERVES WHICH ARE VIRTUALLY CERTAIN (i.e. greater than 90%) to be technically and economically producible

> PROBABLE – those reserves not yet 'proven' but estimated to have better than 50% chance of being technically and economically producible.

> POSSIBLE – those reserves that cannot at present be regarded as 'probable' but are estimated to have a significant but less than 50% chance of being technically and economically producible.

(data from *Development of Oil and Gas Reserves in UK – 1985*)

9.3   Middle East basin

Of the total world proven reserves of oil a large proportion is to be found in the Middle East basin (See Figure 49 and Table 13).

Why is the Middle East so prolific? It is worthwhile finding out because it may hold some clues for future exploration.

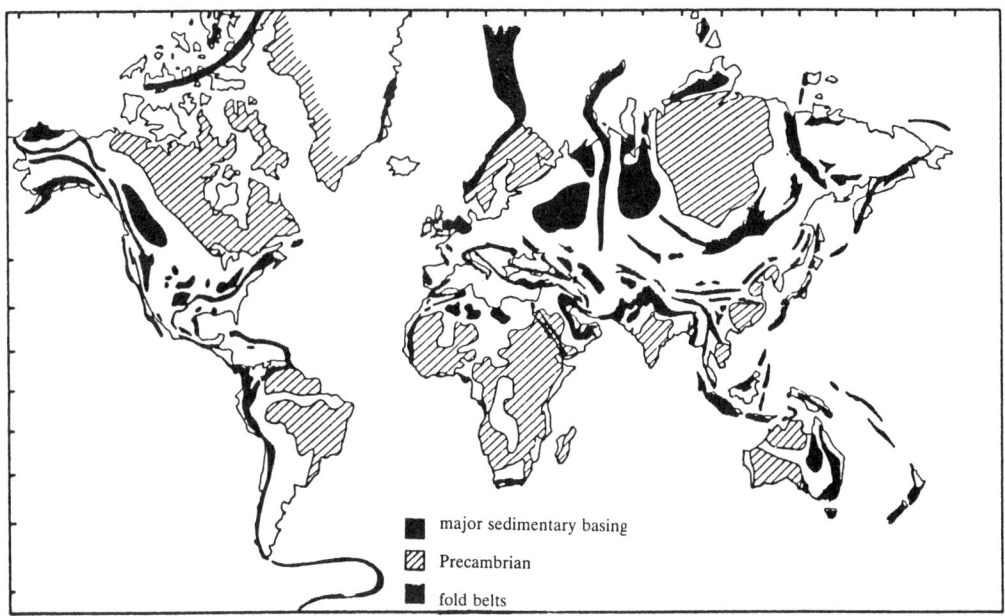

Figure 49 Distribution of the world's major sedimentary basins which contain most of the giant petroleum fields.

Table 13 Economically Recoverable World Oil and Gas Reserves in 1984[*]

Region	World Oil Reserves 10^9 tonnes	% distribution	World Gas Reserves 10^{12} m^3	% distribution
Western Europe	3.1	3.4	4.5	5.0
Eastern Europe	0.3	0.3	0.4	0.4
U.S.S.R.	8.6	9.4	39.6	43.7
Africa	7.6	8.3	5.4	6.0
Middle East	50.6	55.3	21.9	24.2
U.S.A.	3.7	4.1	5.6	6.2
Canada	0.9	1.0	2.6	2.8
Central America	6.8	7.4	2.1	2.3
South America	4.8	5.3	3.1	3.4
P.R. China	2.6	2.8	0.9	1.0
Far East	2.3	2.5	3.8	4.2
Australia/ N. Zealand	0.2	0.2	0.7	0.8
Total	91.5 (approx 670×10^9 barrels)	100.0	90.6	100.0

[*] Data from Decker: World Reserves of Fossil Fuels in 1984.
Glückauf (Translation) 1984 No. 9. p. 158–163.

For most of the Mesozoic and early Cenozoic the African and Eurasian continental plates were separated by a large ocean, called the Tethys ocean. The Middle East basin was part of a broad continental shelf on the southern side of this ocean, and a thick sequence of sediments was deposited on the shelf. When the two plates came together in a *continent—continent collision* in the mid-to-late Cenozoic, the Tethys ocean closed and the sediments were folded. This folding resulted in the formation of the Zagros Mountains at the northern end of the basin (Figure 50).

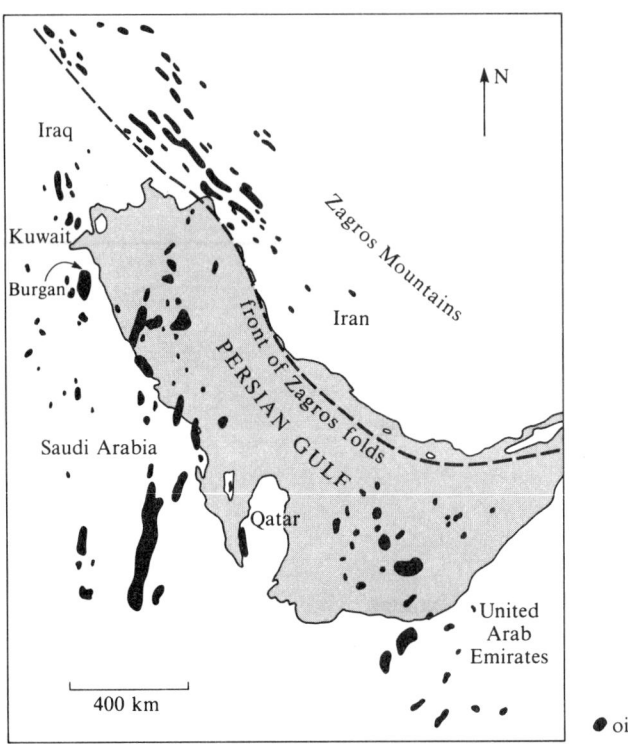

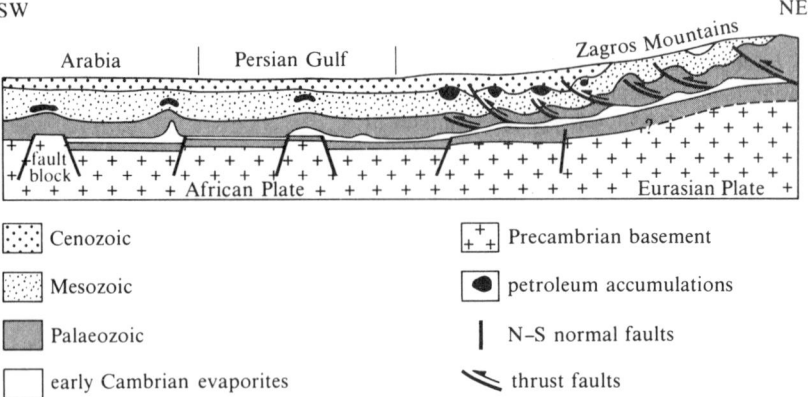

Figure 50 The oil fields of the Middle East basin, and a SW–NE geological section to illustrate the different kinds of trap.

During the Mesozoic–Cenozoic, the Middle East basin was in tropical latitudes, and high biological productivity resulted in an abundance of organic matter in the shallow continental shelf seas. The main petroleum source rocks are organic-rich Mesozoic shales that accumulated in localized depressions on the continental shelf, which was up to 2000 km wide. On other parts of the shelf, shallow-water coarse carbonates and some quartz sands were deposited, and there were interruptions in the deposition of sediments, which resulted in unconformities. Shale deposits are common, and widespread evaporites are well developed near the Jurassic–Cretaceous boundary and later in the Cenozoic.

In the Middle East basin petroleum traps were formed in three ways:

1   Normal north–south faults in the Precambrian basement resulted in the large upstanding fault blocks (Figure 50) over which sediments were draped to form broad anticlines.

2   The early Cambrian salt deposits formed salt plugs which resulted in doming of overlying sediments.

3   The continental collision that caused the Zagros Mountains to form resulted in compressional folds, and in some areas petroleum generated from Mesozoic sediments migrated upwards into the anticlinal traps that were formed.

One of the largest oil fields was discovered in Kuwait in 1983 and is known as the Burgan field (Figure 50); the trap is an elongated dome and its potential was realized because the crest of the anticline exposed at the surface was marked by asphalt seepages.

For the main part the conditions were just right in the Middle East for the production of source and reservoir rocks deep enough for petroleum to be generated.

9.4   World resources of petroleum

There are various ways of estimating the world resources of petroleum. One way would be to add up all the estimated 'recoverable reserves' still left in each known field. This would mean that any future discoveries are not taken into account. It is also possible to estimate the total world resources by the *strata volume method*; this is done by calculating the volume of sediments in the world's basins and multiplying this total by the average yield of oil per unit volume of strata. The USA is the most extensively drilled country, having on average one oil well per 18 km^2; and its total volume of potential oil-bearing strata is 8×10^6 and the estimated total reserves are 165×10^9 barrels.

If it is assumed that all the world's potential oil-bearing sedimentary basins contain the same proportion of oil as in the USA, then another country's resources can be calculated similarly. For example, Canada and the Arctic Islands are thought to have a total of 6×10^6 km^3 of oil-bearing strata. This means the sediments are likely to contain:

$$\frac{165 \times 10^9 \times 6 \times 10^6}{8 \times 10^6} \simeq 124 \times 10^9 \text{ barrels of oil}$$

From calculations of this nature the total world resources were estimated to be 2290×10^9 barrels in 1971, but this is an inflated estimate. An estimate was published in 1984 in the journal *'Glückauf'* of the world's *proven* oil reserves, (details in Table 13) which put them at 670×10^9 barrels. This was taken to be the volume of petroleum in the ground which geological and engineering information indicated to be recoverable oil from known reservoirs under existing economic and operating conditions. This estimate included oil produced by improved recovery methods, which were discussed in the previous chapter. The point to emphasize here is that there is a difference between estimates of resources and estimates of recoverable oil from the proven reserves.

A different method can be used to calculate the world resources of gas, and here we are dealing only with natural gas that is normally associated with oil, sometimes dissolved in the oil or occurring above it in the reservoir and at other times occurring in separate reservoirs. In the USA it has been estimated that there is a ratio of gas to oil of $180\,m^3$ of gas per barrel of oil. Thus, depending on what volume is accepted for world oil resources, an estimate of world gas resources can easily be calculated. If you accept that the proven oil reserves are 670×10^9 barrels, then the gas resources are about $120 \times 10^{12}\,m^3$. This is well above the 1980 *Oil and Gas Journal* estimate of $74.8 \times 10^{12}\,m^3$ of recoverable gas. This figure is increased enormously when dry gas originally derived from coal is also considered, see Table 13.

So what conclusions can we reach regarding world resources of petroleum? Estimates vary considerably, and one of the factors that tends to bring down the figures is the percentage of the oil that it is actually possible to recover from the reservoirs. Even in 1983 less than 50 per cent can be recovered from oil fields, and intensive research is being undertaken to increase this percentage by enhanced recovery methods.

9.5 Output and consumption

Estimates as to how long world reserves of petroleum will last depend on how accurate the estimates of reserves are and on the rate of consumption. In the 1970s it was fashionable to predict that consumption would increase rapidly, but a number of factors have changed this view. North Sea production and UK consumption of oil are shown in Figure 51a; in 1982 the UK produced 30 per cent more oil than it consumed.

World demand for oil fell steadily in the years 1980–82 (Figure 51b) and it fell again in 1983 and 1984 to a level well below the 1979 peak.

Figure 52 shows a series of forecasts of energy demand in the countries of the Organization for Economic Co-operation and Development (OECD)[*] produced by the Paris-based International Energy Agency (IEA). Clearly the estimates of future demand made in 1977 and 1978 have been proved wrong.

What went wrong with these earlier forecasts?

In the decade from the early 1970s the world economics grew by about 20 per cent in real terms, but the energy consumed rose by less than 4 per cent. World industry, the main user of energy, has been hit by a recession in the late 1970s

[*]The OECD is a 26 nation group, based in Paris, which has the prime objective of sustaining economic growth while maintaining financial stability.

and the early 1980s, and this is partly the explanation for the slowing down in energy demand; another reason is drastic energy conservation measures (Chapter 2). The demand for oil has also been affected by a switch to coal and gas.

As far as forecasts of energy demand in the UK are concerned, suffice it to say here that in October 1982 the Department of Energy produced eight different estimates, based on different sets of economic conditions. These estimates ranged from an increase of 33 per cent by the year 2000 to a decrease of 5 per cent. We shall return to the theme of prediction in Part II.

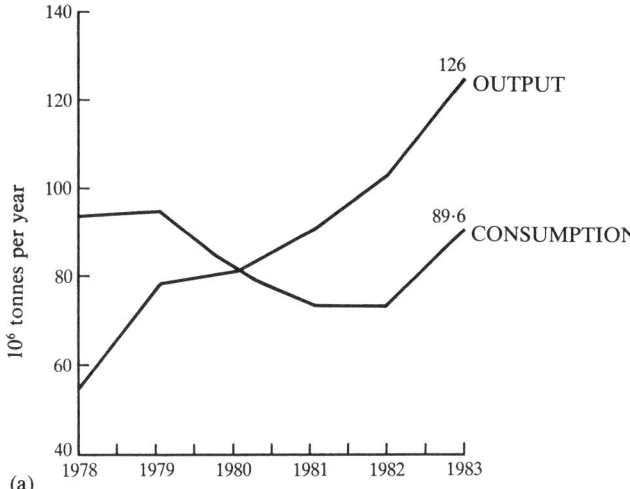

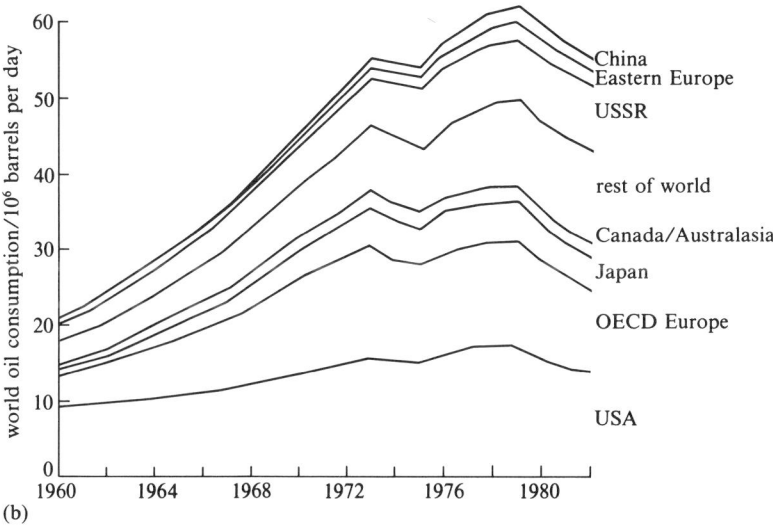

Figure 51 (a) The consumption and output of oil in UK, 1978–83. (b) World oil consumption, 1960–82. (NB One tonne of oil is equivalent to about 7 barrels.)

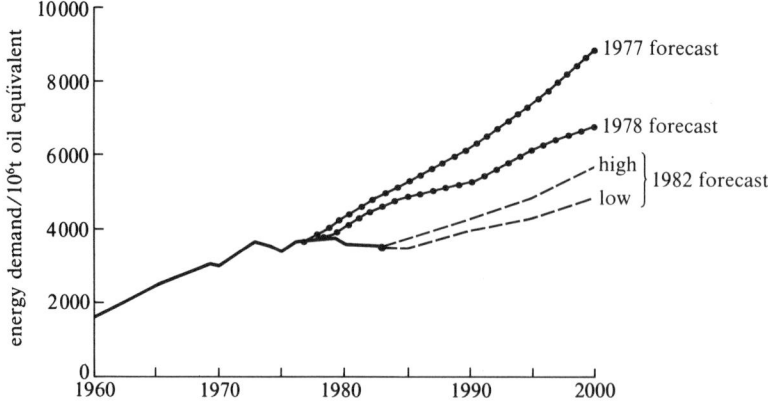

Figure 52   The International Energy Agency forecasts for future world energy demand.

9.6   Coal versus petroleum

Petroleum and coal both originate mainly from organic material that has been subjected to the same geological processes: bacterial action in reducing environments, then burial, causing rocks to be subjected to increased temperature and pressure. But why are the end-products so different? Most coals are remnants of terrestrial plants whereas petroleum is mainly derived from marine phytoplankton, and once generated the oil and gas migrate some distance into porous reservoirs.

Because of its very nature, crude oil and its liquid derivatives can be used in applications where solid coal is useless: the internal combustion engine, for example. However, even some of these machines can use gas, which may be derived from coal, and coal can be converted into petrol.

With the prospects of diminishing petroleum reserves, which are likely to become scarce early in the twenty-first century, technologists are experimenting in producing liquid and gaseous hydrocarbons from coal.

Coal is an attractive feedstock for liquifaction because of the vast reserves. However its composition is different from petroleum in several respects, particularly due to the large numbers of very complex molecules present in coal and its deficiency in hydrogen atoms and excess of oxygen, nitrogen and sulphur. In addition it contains the inerts of mineral matter and moisture.

There are two general routes for the production of liquid hydrocarbons from coal, viz. the synthesis route and the degradation route.

In the synthesis route coal is completely broken down by reaction with steam and oxygen at high temperatures, forming a synthesis gas of carbon monoxide and hydrogen, from which the impurities have been removed. These two gases are then combined in a second stage under the influence of catalysts to produce the required hydrocarbons. The synthesis route is the basis of the South African SASOL plants which have been producing liquid hydrocarbons from coal since 1955. The first gasification stage uses high pressure Lurgi gasifiers and the second stage is based upon the Fischer-Tropsch reactions.

In the degradation route, a selective and less drastic breakdown of the coal structure is achieved in two stages and is potentially the more efficient route.

The first stage involves the solution at 400°C of most of the coal substance in a recyclable solvent rich in hydrogen donor molecules, after which the mineral matter and undissolved coal fractions are removed by filtration. The solution is then reacted with hydrogen at 400°C and 20m. Pa. over selected catalysts similar to those used by the petroleum industry for desulphurisation. These processes have only been operated on the pilot scale as the major unknown is the timescale when coal liquifaction may become economic in the face of declining world petroleum resources.

In conclusion it can be said that as oil and gas reserves diminish, it will become increasingly necessary to direct petroleum to those uses to which it is particularly well suited and avoid using it merely as a source of heat for steam generation in power stations. The best uses for petroleum include liquid fuels for transportation, feedstock for the petrochemical industry and natural gas for domestic heating.

We now leave the hydrocarbon fuels and in the second part of this book turn our attention to other forms of energy resources.

9.7   Summary of Chapter 9

1   A number of small gas and oil fields have been discovered onshore in Britain. In the East Midlands, for example, there are traps in the Carboniferous, and oil has been discovered in the Triassic and Jurassic rocks in southern England. The oil and gas fields in the North Sea are mainly in Permian (gas) and Jurassic and early Tertiary (oil) reservoirs; the source of the gas in the southern North Sea basin is the Carboniferous coals, and the Jurassic Kimmeridge shales are the source rocks for oil and gas in the north North Sea basin.

2   The UK reserves in the North Sea are minute compared with those of the Middle East which has 55 per cent of the world proven reserves of crude oil, amounting to 51×10^9 tonnes. Ideal conditons prevailed in the Middle East during the Mesozoic—Cenozoic for the production of enormous volumes of source rocks which underwent maturation after the formation of numerous large traps.

3   There are various ways of estimating world resources of petroleum; one is by calculating the volume of oil-bearing sediments in the world's sedimentary basins and multiplying this by an average yield of oil per unit volume. Estimates of the total world resources peaked in 1971 at 2290×10^9 barrels; a figure of 670×10^9 barrels of recoverable oil was published in 1984. An estimate of gas resources is $120 \times 10^{12} m^3$, and recoverable gas is estimated at $90.6 \times 10^{12} m^3$.

4   Two factors that may prolong the "life" of petroleum reserves are the slowing down in consumption during the recession of the 1980s and steps being taken to increase the efficiency of energy utilisation, described in Part II.

Further Reading for Part I

1. The Open University, 1984, *S238: The Earth's Physical Resources*. OU Press, Milton Keynes.
2. C R Ward (ed.), 1984. *Coal Geology and Coal Technology*. Blackwell Scientific Publications, London.
3. P J Adams, 1960. *Origin and Evolution of Coal*. H M S O, London.
4. Assessment of Energy Resources, 1981. Report No. 9, The Watt Committee on Energy, Ltd. London.
5. Coal and the Environment, 1981. Report of Government Standing Committee, Department of Energy, London.
6. Development of oil and gas resources of UK 1985. Dept of Energy, London.
7. J Brooks (ed.), 1981. *Organic Maturation Studies and Fossil Fuel Exploration*. Academic Press, London.
8. G Foley, 1981 (second ed.) *The Energy Question*. Penguin, Harmondsworth, London.

PART II

1 Nuclear power

Study Comment. In this Chapter we see how natural fission of some radioactive isotopes (especially uranium) can be artificially enhanced in nuclear reactors. The distribution, extraction and reserves of uranium ores are reviewed in relation to the future potential of the nuclear industry.

1.1 Nuclear reactions and reactors

The *nuclei* of heavy isotopes of elements such as uranium and thorium experience *radioactive decay*, with the emission of *alpha particles*. In addition, the nuclei of these isotopes undergo *fission* into two smaller nuclei of roughly equal mass, along with some very light particles (*neutrons*, in particular), and a good deal of energy. The most abundant isotope of uranium is uranium-238 (^{238}U), with a nuclear mass 238 times that of a single proton or neutron. In the natural state ^{238}U will undergo one fission for roughly every million radioactive alpha decays. To see why this rate of fission must be increased in order to run a nuclear power station, consider the following calculation.

The natural radioactive heat production of ^{238}U is $3\,000$ K kg^{-1} yr^{-1} so, in other words, a 1 kg mass of this isotope will produce 3×10^3 J in one year through natural spontaneous decay. Now a 1 kg mass of coal, as you know from Part I of this book, produces an energy output of 2.8×10^7 J. So it will take $2.8 \times 10^7/3 \times 10^3 = ca.\ 10^4$ years (10,000 years) for spontaneous radioactive decay to produce the energy equivalent of the same mass of coal. Clearly, for the efficient operation of nuclear reactor power stations, a much more rapid method of obtaining the energy stored in uranium is required.

In a nuclear reactor, uranium atoms are bombarded with neutrons; this encourages them to realease their stored energy much more rapidly. The reaction, which forms the basis of most nuclear power generating plants in existence today, involves the capture of low energy neutrons, called *slow neutrons* by nuclei of the less common uranium isotope ^{235}U. This capture of neutrons increases the nuclear mass by one unit, producing ^{236}U:

$$^{235}_{92}U + ^{1}_{0}n \rightarrow ^{236}_{92}U \qquad (1)$$

This isotope is highly unstable and breaks down in the presence of slow neutrons, releasing energy:

$$^{236}_{92}U \rightarrow ^{95}_{39}Y + ^{139}_{53}I + 2^{1}_{0}n + 3.2 \times 10^{-11} \text{ J} \qquad (2)$$

unstable   typical   neutrons   energy
fission
products

There are two important things to note about this reaction: first, the main fission products (yttrium-95 and iodine-139) are radioactive and will undergo further decay, releasing more energy, until they reach stability; and second, more neutrons are produced.

What will happen if these neutrons collide with more ^{235}U nuclei?

The reaction will continue, producing more neutrons and more energy in an uncontrolled *chain reaction*, of the kind that occurs in atomic bombs, unless it is controlled in some way, as in a power station.

Equations 1 and 2 allow us to calculate the amount of energy produced by uncontrolled fission, known as the *fission energy* of uranium, as follows: the *relative atomic mass* of ^{235}U is, of course, 235 and so 1 kg of ^{235}U will contain $1\,000/235 = 4.25$ moles. There are 6×10^{23} atoms (the *Avogradro constant*) in 1 mole, and so 1 kg contains $4.25 \times 6 \times 10^{23}$, or 25.5×10^{23} atoms. The fission energy available from one atom is given in equation 2, and so the amount available from 1 kg will be $(25.5 \times 10^{23}) \times (3.2 \times 10^{-11}) \approx 8.2 \times 10^{13}$ J. Several points follow from this calculation.

> How does the fission energy available from 1 kg of ^{235}U compare with the annual radioactive output of 1 kg of ^{238}U and with the combustion energy from 1 kg of coal?

1   You should realize that the fission of 1 kg of ^{235}U produces 3×10^{10} times more energy than is produced in one year by 1 kg of ^{238}U undergoing radioactive decay. Of course, fission energy per unit mass is not very different in *total* from that which is ultimately released by radioactive decay — it is merely released all at once instead of over thousands of millions of years.

2   The fission energy from 1 kg of ^{235}U is about 3×10^6 times the amount of energy obtained from 1 kg of coal. So, weight for weight, ^{235}U is three million times more concentrated than coal as an energy source — and weight for weight it costs only about ten thousand times more than coal to mine, process and transport to the power station.

3   Natural uranium contains much less ^{235}U than ^{238}U — in fact only 0.7 per cent is ^{235}U, and 99.3 per cent is ^{238}U. This means that only 0.7 per cent of the fission energy calculated above will be obtained by the reactions in equations 1 and 2 from a 1 kg block of natural uranium. For this reason, other relatively new power stations, with increased technical complexity, have been designed to extract fission energy from the more abundant and more stable isotope ^{238}U, according to the following reaction:

$$^{238}_{92}U + ^{1}_{0}n \rightarrow ^{239}_{92}U \rightarrow ^{239}_{94}Pu \rightarrow ^{100}_{40}Zr + ^{137}_{54}Xe + 2^{1}_{0}n + 3.4 \times 10^{-11} \text{ J} \qquad (3)$$

	rapid unstable	typical fission neutrons energy
	β-decay	products,
	(see below)	zirconium and
		xenon

One reason why reactors using this process depend on a more advanced technology is that a beam of highly accelerated, high energy *fast neutrons* is required to establish an efficient chain reaction. The nuclei of $^{239}_{92}U$ which are produced undergo rapid β-*decay* (in which neutrons are converted into protons) in producing $^{239}_{94}Pu$, and it is this plutonium that is readily fissile in the presence of fast neutrons. Clearly, by exploiting the more abundant uranium isotope in this way, the potential of naturally occurring uranium as an energy resource can be increased by almost 150 times (100/0.7) compared with the energy extractable from ^{235}U alone (as the amounts of energy released in equations 2 and 3 are similar).

1.1.1   Types of nuclear reactor and fuel requirements

There are two main groups of nuclear reactor depending on whether the reaction in equation 2 or 3 is exploited: burner reactors and fast breeder reactors.

Burner reactors

Here the main fuel is ^{235}U and the object is to use up or 'burn' as much of it as possible. Because natural uranium contains only 0.7 per cent ^{235}U, *enriched uranium*, containing up to 3 per cent ^{235}U, is manufactured to increase the efficiency of some types of burner reactor. In order to perpetuate the chain reaction in equations 1 and 2, which results in neutrons with a wide range of energies, fast neutrons are slowed down using a *moderator* and the *reaction rate* is adjusted using *control rods* that totally absorb neutrons (Figure 1). The rods can be raised or lowered to increase or decrease the heat output. The moderator may be graphite or heavy water (deuterium oxide, ^{2}H$_2$O) whilst the control rods are usually made of boron. Even in this type of reactor, neutrons are absorbed by a small proportion of the ^{238}U in the fuel rods to produce ^{239}Pu, nuclei of which may then undergo fission and add appreciably to the total energy released. But the amount of plutonium created is much less than the amount of uranium used up and this is another reason why such reactors are called 'burner reactors'.

Fast breeder reactors

If the neutrons produced in the chain reaction are not moderated (i.e. some are fast neutrons) then the rate of conversion of ^{238}U into ^{239}Pu can be made to exceed the fission rate of ^{239}Pu. Reactors using fast neutrons to create more ^{239}Pu fuel than they use are known as fast breeder reactors. The main fuel is ^{238}U, together with an initial charge of ^{239}Pu which is needed to set in progress the chain reaction which will use fast neutrons to 'breed' ^{239}Pu from ^{238}U continuously during the lifetime of the reactor. The initial charge may be extracted from the spent fuel rods of burner reactors or, as will increasingly be the case, from the excess ^{239}Pu produced in breeders (where the rate of plutonium production exceeds its rate of fission). In practice, about 60 per cent of the fuel elements in breeder reactors will be converted into useful energy, compared with 0.5–1.0 per cent in burner reactors. The breeder process looks very much like getting something for nothing, and it is true that fast breeders could greatly extend the lifetime of nuclear fuel reserves (Section 1.3).

Figure 1 shows the essential design features of four main types of reactor. Three of these (Magnox, AGR and PWR) are the most common types of burner reactor in use today; there are relatively few fast breeder reactors. These four types of reactor have certain features in common: they all have *cores* encased in steel vessels surrounded by concrete shields. Moreover, as with fossil fuel power stations, the heat generated is used to convert water into steam which is then directed under pressure into the generator turbines after which it is condensed and returned to the boilers. Whereas in fossil fuel stations the hot gaseous combustion products of coal or gas heat the water in the boilers directly, in nuclear power stations there is no combustion, and heat is transmitted to the boilers by a sealed-in coolant fluid which circulates between the boilers and the hot reactor core: the steam boiler acts as a heat exchanger. There are three stages involved in transferring heat away from the reactor core: (i) the core coolant circuit which uses carbon dioxide gas, water or sodium depending on the reactor type (Figure 1); (ii) the heat from stage (i) is transferred to a closed steam-water circuit, which includes the turbine plant; (iii) the steam in circuit (ii) is condensed to water on the low-pressure side of the turbine by a cooling water circuit that is open to the environment. The requirement for large volumes of cooling water means that nuclear reactors must be sited in coastal locations or near large rivers.

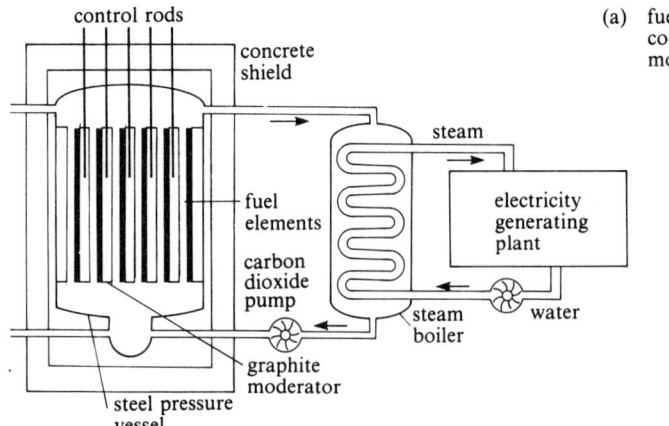

(a) fuel: natural metallic uranium
coolant: carbon dioxide gas
moderator: graphite

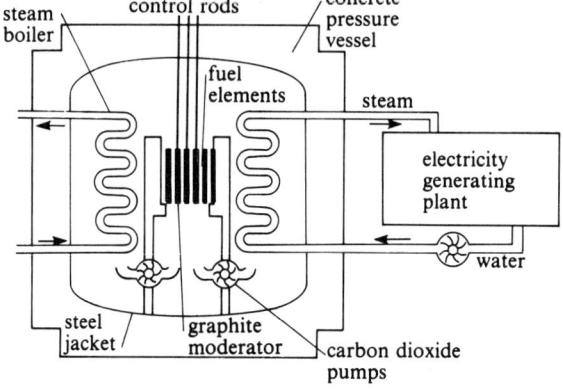

(b) fuel: slightly enriched
uranium oxide
coolant: carbon dioxide gas
moderator: graphite

(c) fuel: enriched uranium oxide
coolant: water
moderator: water

Figure 1 Diagrams of the essential features of (a) a Magnox reactor, (b) an advanced gas-cooled reactor, (c) a pressurized water reactor, and (d) the fast breeder reactor system now being developed for commercial purposes.

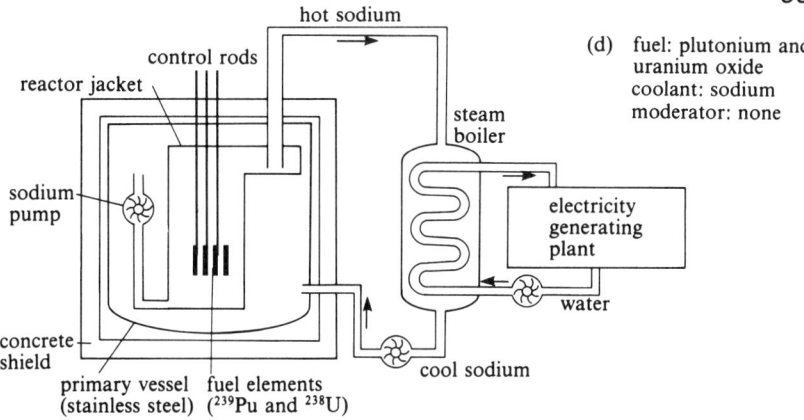

hot sodium

control rods

reactor jacket

(d) fuel: plutonium and
uranium oxide
coolant: sodium
moderator: none

steam boiler

sodium pump

electricity generating plant

water

concrete shield

cool sodium

primary vessel   fuel elements
(stainless steel) (^{239}Pu and ^{238}U)

Magnox reactors (Figure 1a) are fuelled by metallic uranium containing the natural proportion of ^{235}U (i.e. 0.7 per cent) held in tubes of a magnesium alloy (Magnox). The moderator is graphite, the coolant is carbon dioxide and the operating temperature is about 400°C.

Advanced gas-cooled reactors (AGRs; Figure 1b) are very similar to Magnox reactors, except that uranium oxide, enriched in ^{235}U (2.3 per cent instead of 0.7 per cent) and packed in stainless steel tubes, is used as a fuel. The moderator and coolant are as in Magnox reactors but the operating temperature, 800°C, is higher, leading to greater efficiency and output. Although the capital cost of building AGRs is very high, they have many additional safety features over Magnox reactors and, because of their greater efficiency, produce cheaper electricity (Section 1.1.2).

Pressurized water reactors (PWRs; Figure 1c) use enriched uranium oxide fuel rods in zirconium alloy tubes, have pressurized water as both coolant and moderator, and operate at 300–400°C. PWRs came under serious criticism following the Three Mile Island incident in Pennsylvania during 1979, when safeguard mechanisms failed to prevent a serious increase in the temperature and pressure of the reactor core, followed by a leak of radioactive materials. The reactor developed a gas bubble in its core, increasing the danger of further rise in temperature and ultimately a 'meltdown' which would involve the melting of the reactor core containment vessel. Although some subsequent reports have indicated that the damage arose from technical failures rather than from fundamental design weaknesses, there is still considerable public opposition to PWRs (as evidenced by the Sizewell enquiry, in the UK, which started in 1983 and which continues at the time of writing (1986) — see Section 1.1.2).

Fast breeder reactors (Figure 1d) use the available uranium fuel so much more efficiently than burners that they would seem to be the obvious choice for power generation in the future. The intense heat generated by the fast neutron chain reaction requires the circulation of highly conductive liquid sodium as the coolant, and it is this aspect that presents formidable technical problems, especially as the operating temperature is around 600°C.

Figure 1d shows that there is no moderator in a fast breeder reactor: why is this?

Remember that the breeder reacton (equation 3) depends on the ability of fast neutrons to transmute ^{238}U into ^{239}Pu and to sustain the fission reaction of plutonium. Because the moderator in burner reactors is designed to *slow down* the fast neutrons produced in the chain reaction, it is out of place in a breeder reactor.

The efficient operation of the reactor also requires a much higher density of fast neutrons than that of slow neutrons required in burner reactors. The neutrons emerging from the chain reaction must not make collisions that would slow them down before they strike ^{238}U nuclei; hence the core region of a fast breeder must be more compact than that of a burner reactor and, besides containing no moderator, it must contain only the minimum volume of coolant.

The state of development of the nuclear power industry in the world in mid-1983 is summarized in Table 1; notice the dominance of North American and European countries together with the USSR and Japan as major nuclear power consumers. Of the burner reactors operating, just over a hundred were PWRs and many others were related types of water-cooled design not shown in Figure 1. Notice also the small number of breeder reactors: despite the formidable technological difficulties mentioned earlier, however, there are prototype fast breeder reactors operating in several countries. The first commercial fast breeder is now being built in France.

Table 1 Distribution of operational nuclear reactors in mid 1983.

Country	Burners	prototype breeders
USA	78	1
Britain	32	1
USSR	31	3
Japan	26	—
West Germany	18	1
France	30	2
Canada	12	—
Scandinavia	14	—
Italy	3	—
Spain	5	—
India/Pakistan	5	—
Argentina	1	—
Korea and Taiwan	6	—
Other European countries	26	1
totals	287	9

Turning now to consider the contrasting fuel requirements of burner and breeder reactors, bearing in mind that the fission energy of ^{235}U is 8.2×10^{13} J kg^{-1} and that of ^{238}U is 8.6×10^{13} J kg^{-1} (from equations 2 and 3), we need to consider:

1 The thermal efficiency of the reactor: for modern burner reactors, this is about 35 per cent (i.e. 65 per cent of the available energy is lost within the reactor—coolant—turbine system);

2 The power output: a typical value for a burner reactor is 1 000 MW (3.2×10^{16} J yr^{-1});

3 The lifetime of the reactor: say 30 years.

The question then is what mass of naturally occurring uranium (0.7 per cent ^{235}U) would need to be processed to provide the fuel elements for a 35 per cent efficient, 1 000 MW reactor over its 30 year lifetime. Clearly, the total power output over 30 years is:–

$$3.2 \times 10^{16} \times 30 = 9.6 \times 10^{17} \text{ J}$$

But the reactor has a thermal efficiency of 35 per cent, and so the energy 'input' must be 100/35 times the energy output; moreover, since 1 kg of ^{235}U produces 8.2×10^{13} J, the amount of ^{235}U required is:

$$\frac{9.6 \times 10^{17} \times 100/35}{8.2 \times 10^{13}} = 33\ 500\,\text{kg}$$

Natural uranium contains only 0.7 per cent ^{235}U, and so the total mass of uranium required will be:

$$\frac{33\ 500 \times 100}{0.7} = 4.8 \times 10^6\,\text{kg} = 4800 \text{ tonnes}$$

This means that the average annual fuel requirements of the 287 burner reactors identified in Table 1 which have an average output of 750 MW is:

$$\frac{4800 \times 287 \times 0.75}{30} = 34\ 400 \text{ tonnes}$$

We shall look further at uranium fuel supply and demand in Section 1.3.

Fuel elements do not last for 30 years, and in order to maintain peak performance, a proportion (about 15–20 per cent) of the rods are changed each year. Of the fuel requirements calculated above, a much larger amount, about 600 tonnes, is required as an initial fuelling charge than would be required annually to replace spent fuel rods, an average of 140 tonnes a year. As noted above, the spent fuel rods will comprise some 96 per cent of *depleted uranium*, which has lost most of its ^{235}U component (about 4600 tonnes), and under 1 per cent of ^{239}Pu (up to 50 tonnes depending on reactor characteristics), both of which are potentially available for a fast breeder cycle. The remaining 2–3 per cent of a spent fuel rod is mainly composed of fission products. The fuel requirements and products of burner reactors are summarized in Figure 2a.

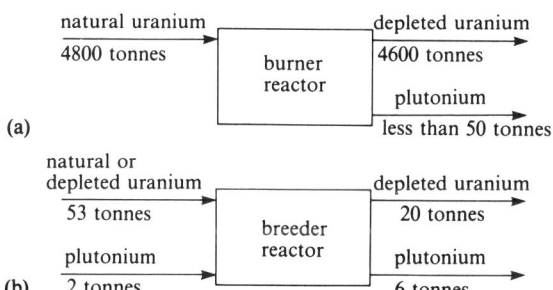

(a)

(b)

Figure 2 A summary of the fuel requirements and spent fuel products from (a) a typical burner reactor and (b) a fast breeder reactor, based, in both cases, on a power output of 1000 MW for 30 years.

The next step is to quantify the fuel requirements for fast breeder reactors and, again, we will consider a 1000 MW reactor operating for 30 years. A probable efficiency for breeder reactors is also 35 per cent but in this case much more of the fuel, estimated at 60 per cent of that available, will be used.

On the same basis as before, the energy output of a 1000 MW breeder reactor is 9.6×10^{17} J over 30 years. With a thermal efficiency of 35 per cent and a ^{238}U energy content of 8.6×10^{13} J kg^{-1}, the amount of fuel required is:

$$\frac{9.6 \times 10^{17} \times 100/35}{8.6 \times 10^{13}} = 32\,000 \text{ kg}$$

If only 60 per cent of the available atoms undergo fission, the total mass of uranium required will be:

$$\frac{32\,000 \times 100}{60} = 53\,000 \text{ kg} = 53 \text{ tonnes}$$

Given that an initial plutonium charge of about 2 tonnes is required, the total fuel load for a breeder reactor producing the same amount of electrical power over the same period as the burner reactor (above) is nearly 100 times smaller (Figure 2b). If you were not convinced before, this should really emphasize the point that burner reactors are extravagantly wasteful of nuclear fuel — unless the depleted uranium can be reprocessed and used later in a breeder reactor. The final point to notice from Figure 2b is that a typical breeder reactor will contain more ^{239}Pu in its fuel rods after 30 years than is invested in the initial charge. Unlike burner reactors which must be serviced periodically with undepleted fuel rods during their active life, breeder reactors require relatively little attention once they have received their initial charge. Figure 3 illustrates, in a light-hearted way, the annual supply of fuel required to service 1000 MW power stations using coal, burner and breeder technologies once they are in production.

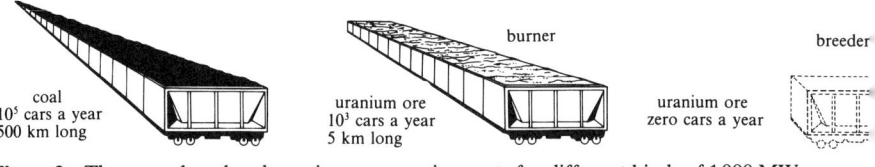

coal
10^5 cars a year
500 km long

uranium ore
10^3 cars a year
5 km long

burner

uranium ore
zero cars a year

breeder

Figure 3 The annual coal and uranium ore requirements for different kinds of 1000 MW power station. (An average grade of 0.3 per cent uranium in the uranium ore was assumed for this figure.)

1.1.2 Nuclear power in Britain

Much of the work that paved the way for nuclear power generation took place in British universities during the early years of the twentieth century. During World War II, however, scientific effort was concentrated in the USA on the military uses of nuclear energy, though the first nuclear power reactor was built in Canada at this time. Research reactors were established in the UK during the late 1940s, and the first commercial power station, a Magnox reactor, was opened at Calder Hall (Figure 4) in 1956. The development of the industry has since been overseen on behalf of the Government by the United Kingdom Atomic Energy Authority (UKAEA) together with its subsidiary British Nuclear Fuels Ltd (BNFL), which is responsible for fuels, and the National Nuclear Corporation (NNC), which is responsible for the design and construction of commercial reactors. The commercial stations are owned and operated by the Central Electricity Generating Board (CEGB) in England and Wales and, further north, by the South of Scotland Electricity Board (SSEB).

Nine of the commercial stations shown in Figure 4 are improved versions of the Calder Hall Magnox reactor. They were opened during the period 1962–71 and have generating capacities between 245 and 840 MW (average 500 MW = 1.6 ×

10^{16} J yr^{-1}). The five most recent commercial stations are AGRs, opened between 1976 and 1982, with generating capacities of about 1000 MW (3.2 × 10^{16} J yr^{-1}). Two more AGRs being built, Heysham 2 and Torness, are due to be commissioned in the late 1980s. The CEGB's plans to build Britain's first

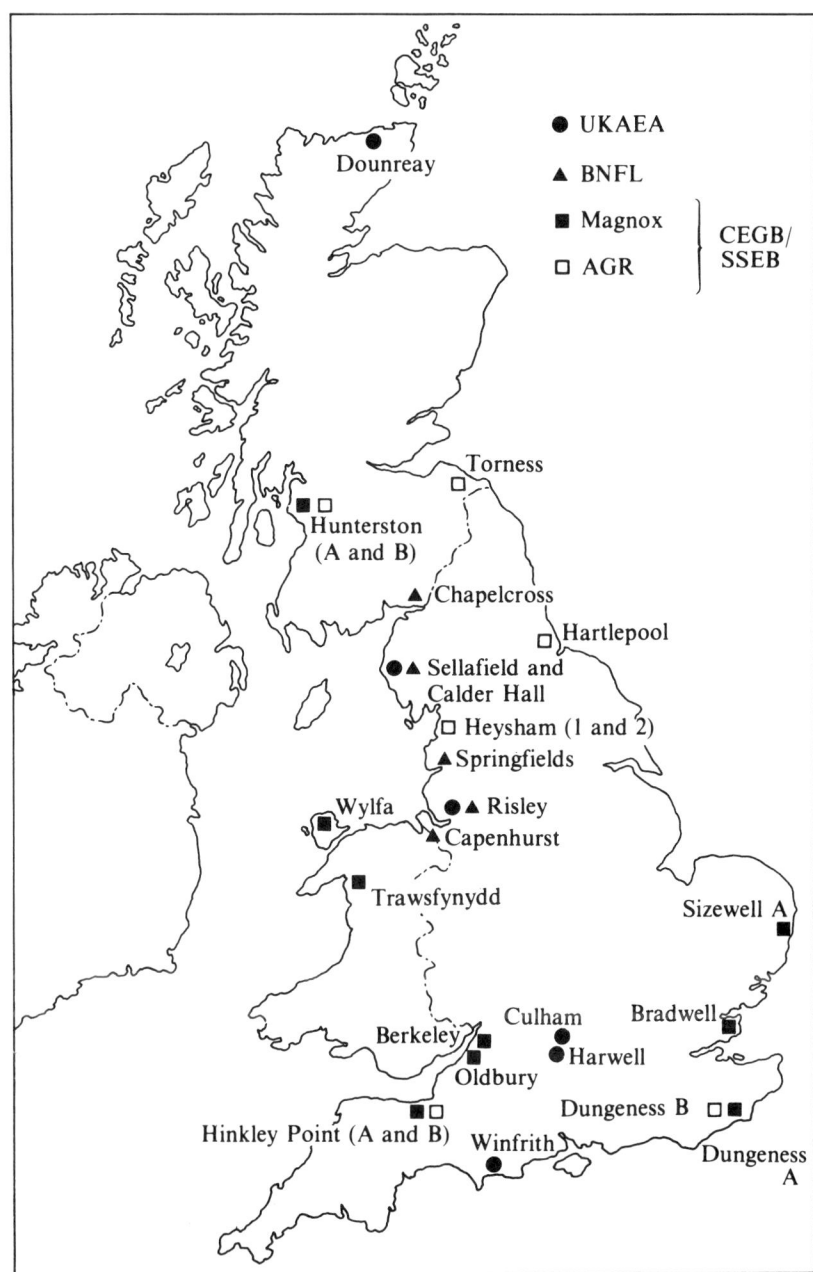

Figure 4 The most important nuclear research establishments (circles and triangles) and commercial power stations (squares) in Britain in 1983. Several small research reactors are not shown. Reactor centres owned and operated by the UKAEA and BNFL are distinguished from the commercial CEGB/SEB sites.

PWR in the 1990s as Sizewell B were being debated hotly at the time of writing, the objectors arguing that a further nuclear power station will not be required on this timescale and, more important, emphasizing the suspect safety record of PWRs. The strength of the CEGB's argument lies in proving that the design and safety targets for a British reactor would be acceptable and that nuclear power is cheaper to produce than coal-fired power. However, this claim of cost-effectiveness depends critically on the accounting technique. Purely in terms of fuel and running costs, it costs 1.45p to produce a unit of electricity (1 kWh) from an AGR in 1983 compared with 1.87p for a unit of coal-fired electricity. However, if the initial costs of building a nuclear reactor (estimated at £2000 million for Sizewell B) are included and discounted over a probable lifetime of 30 years, this adds almost 1p a unit to nuclear-powered electricity compared with 0.2 to 0.3p a unit for the capital costs (about £500 million) for the equivalent output from coal-fired plant. Thus the price advantage of nuclear power is wiped out or even reversed.

Including contributions of 250 MW from the prototype fast breeder reactor at Dounreay (Figure 5) and of lesser amounts from other reactors at the UKAEA/BNFL research establishments (Figure 4), the electricity generating capability of British nuclear reactors was about 10000 MW in 1982, or 15 per cent of the total output capacity of British power stations. In practice, the actual output from nuclear stations was about 20 per cent of the total output as, because of their low operating costs compared with fossil fuel stations, they are used in preference to other sources of power which are added as and when required.

Figure 5 The Dounreay prototype fast breeder reactor on the north Scottish coast, which has been operating since 1975 (foreground); the experimental fast breeder core, built in 1959, lies within the white dome in the background.

Spent fuel rods from burner reactors contain mainly ^{238}U and its oxide, together with ^{239}Pu (less than 1 per cent) and radioactive fission products (about 2 per cent). As their number has increased, there has been a growing need to organize large scale *reprocessing* of these spent fuel rods and to consider practices for the safe disposal of nuclear wastes. At Sellafield (Windscale), BNFL reprocess fuel rods to recover ^{238}U and ^{239}Pu, which can be converted into fuel for breeder reactors or stored for future use. This leaves the radioactive wastes, which are separated according to their level of radioactivity into 'low', 'intermediate' and 'high' level wastes for disposal or long-term storage (Section 4.2). The extension to the Sellafield site, which was the subject of a public inquiry in 1977, concerned a plant for the reprocessing of oxide wastes (*THORP* = thermal oxide reprocessing plant). Planning permission was granted and the THORP plant, which should be completed before 1990, will have a capacity to reprocess 1200 tonnes a year of spent oxide fuel. About half of this capacity will be required for the output of British reactors, and BNFL signed contracts in 1978 with Japanese and European electricity companies in order to exploit the remaining capacity.

1.2 Uranium geology

Study Comment. The next stage is to consider the availability of uranium resources and then, in Section 1.3, to enquire what contribution nuclear reactors may make to future energy demand and at what cost in monetary and environmental terms. We shall use the data from Section 1.1 and Figure 2, after examining the geology of uranium which provides a basis for evaluating the supply and demand situation.

Our starting point in the search for economic deposits of uranium-rich rocks that may be suitable for mining, concentration and exploitation as nuclear fuels, is to remind you that uranium is an element that tends to occur preferentially as oxide minerals in granitoid igneous rocks. The average chemical composition of the upper continental crust is close to *granodiorite* which contains, on average, 3 p.p.m. uranium. But would a typical continental granodiorite constitute an economic source of uranium?

At a concentration of 3 p.p.m. (i.e. 3g per tonne of material), the 4800 tonnes of uranium required to keep a single 1000 MW burner reactor in production for 30 years would come from:–

$$\frac{4800 \times 10^6}{3} = 1.6 \times 10^9 \text{ tonnes of granodiorite}$$

At a density of 2.7 tonnes m^{-3}, this granodiorite will occupy *ca.* $600 \times 10^6 \text{ m}^3$, and thus be produced from the equivalent of a cubic-shaped hole of size about 840m, assuming 100 per cent uranium extraction. Even then, 999.997 kg of every tonne of rock mined will be waste; clearly, such low grades can hardly be described as economic uranium ores.

Obviously we must look for rocks in which this element is more concentrated than in average granodiorite. *Uranium ore deposits* containing upwards of 350 p.p.m. uranium (0.035 per cent) are being exploited, though the minimum concentration, or *cut-off grade* being mined is subject to local economic conditions and, for some operations, is considerably higher than 350 p.p.m.

But what processes go to form these uranium ores and what kinds of deposit are involved? To answer these questions, we need to develop some concepts about the variety of geological settings in which uranium occurs. Uranium ores can be assigned on the basis of their geological setting into the following six categories of ore types, which we shall be considering separately:

1 *Pegmatites* and disseminated magmatic uranium deposits. *Disseminated deposits* are those containing dispersed, discrete uranium-rich particles which, on average, are economic to mine.

2 *Hydrothermal vein deposits* in cracks, fissures, *stockworks* and fault zones.

3 Hydrothermal vein deposits developed beneath ancient (Proterozoic) *unconformities*; these are known as *vein–unconformity deposits*.

4 Sandstone deposits related to sediments in rivers, lakes and shallow-water seas.

5 *Placer deposits* in quartz-pebble *conglomerates*.

6 Other sedimentary concentrations in phosphatized limestones, marine phosphates and black shales.

Of these six classes, types 1–3 are formed by igneous/hydrothermal processes (perhaps with a component of surface processes in type 3) whereas types 4–6 are deposits formed by surface processes. Types 1–3, formed by internal processes, are usually termed *primary uranium deposits*, whereas types 4–6 are *secondary uranium deposits* because a prerequisite for their formation is that primary sources of uranium (not necessarily of ore grade) must be recycled in surface environments. In Table 2 we summarize the range of uranium ore grade, the associated metals, the largest known deposit and the contribution of each type of deposit to uranium reserves as a prelude to taking a more detailed look at uranium deposits. But, first, to understand the formation of uranium deposits we need to consider a few aspects of uranium geochemistry.

Table 2 Summary of the main categories of uranium ore deposits

Type	Uranium grade (%)	Commonly associated metal(s)	Largest known deposit/tonnes of uranium	Proportion of world-wide reasonably assured resources (%)*
pegmatites and disseminated magmatic deposits	0.03–0.13	tin, tungsten	100 000	14
vein-type deposits	0.1–2.0	molybdenum	20 000	5.5
vein-unconformity deposits	0.4–4.0	nickel	200 000	16
uranium in sandstones	0.05–0.5	vanadium	40 000	39
quartz-pebble conglomerates	0.01–0.1	gold	15 000	16.5
others†:				
calcrete	about 0.15	—	40 000	
limestones	0.2–0.3	—	10 000	9
phosphates	0.001–0.06	phosphorus	10 000	
black shales	0.001–0.03	iron	500 000	

* This term is explained fully in Section 1.3; reasonably assured resources are all the reserves and conditional resources valued at less than 130 US dollars per kilogram of uranium in 1984 prices.

† These types of deposit are described and defined in Section 1.2.6.

1.2.1 The geochemistry of uranium

There are two aspects of importance: the first is concerned with the *ionic size* and *charge* of uranium, which influence its behaviour in *magma*; the second relates to its solubility in water and determines its role in surface environments.

Electronegativity influences the kinds of ore mineral that will form; elements with low electronegativity (less than about 1.8) will form oxides. Uranium has an electronegativity of 1.7 and occurs typically as oxides such as uraninite (U_3O_8, or more corectly $UO_2U_2O_6$ in which $UO_2 > U_2O_6$). *Accessory minerals* rich in uranium also occur in igneous rocks. But why does uranium not enter the crystalline structures of the major silicate minerals? To answer this we need to know that uranium forms two ionic species U^{4+} (ironic radius 97 pm) and U^{6+} (ionic radius 80 pm). Both species are much larger than Si^{4+} (ionic radius 42 pm), which is a component of the *polyanions* of silicate minerals, and both are more highly charged than the common cations (Fe^{2+}, Mg^{2+}, Ca^{2+}, Na^+, etc.) that link polyanions together to form mineral structures. This means that uranium will not readily be taken up by the major silicate minerals that form over a range of temperatures during the crystallization of a silicate magma, but will be progressively enriched in the residual magma. Uranium forms its own oxide minerals, or substitutes in other accessory oxides (e.g. for Ca^{2+}, radius 99 pm, in apatite — $Ca_5(PO_4)_3(OH,F)$; or for Zr^{4+}, radius 79 pm, in zircon — $ZrSiO_4$) that occur in the residual and usually most silica-rich parts of slowly cooled intrusions, or in the watery, silicate-rich fluids associated with pegmatite formation. Although these minerals may be forming throughout the cooling history of an igneous melt, they are generally most abundant in the late stages of crystallization.

So the large charge and size of uranium cations influence their behaviour in silicate magmas, but it is the relative solubility of the two cations that determines their behaviour in surface environments. Figure 6 is a simplified version of the Eh–pH diagram for uranium. Imagine that a "primary" uranium deposit, such as a pegmatite containing uraninite, is being eroded. In the surface environment,

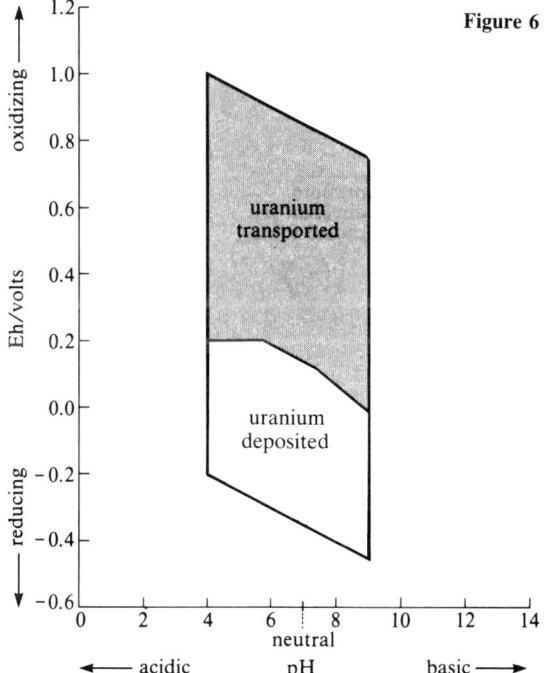

Figure 6   Simplified Eh-pH diagram for uranium.

as this uraninite moves from the site of weathering to a zone of deposition, it will remain undissolved *only* if conditions are reducing (i.e. low Eh) and this is not significantly affected by variations in pH. In contrast, uraninite breaks down to be transported in solution when conditions are relatively oxidizing (high Eh). So the behaviour of uranium in surface environments is governed by the fact that the more oxidized form, U^{6+}, is highly soluble whereas the relatively reduced form, U^{4+}, is highly insoluble. When uranium-bearing rocks are weathered in oxidizing conditions, any U^{4+} is rapidly oxidized to U^{6+}, and most of the uranium in the weathered rock is therefore dissolved. It is then carried as soluble compounds by surface water in rivers, or by sub-surface groundwaters, until reducing conditions are encountered and precipitation occurs.

1.2.2 Pegmatites and disseminated magmatic deposits

These uranium deposits originate by magmatic process whereby uranium is selectively concentrated in the residual liquids of partially crystallized magmas that cooled over a long time interval. Such residual liquids usually crystallize to form the *most* granitic parts of large intrusions where rock types may vary from gabbro to granite (e.g. the Bushvelt intrusion of South Africa). But the uranium concentration in most granites lies between 2 and 20 p.p.m. So only exceptional granites contain economic concentrations of uranium. How might these exceptional granites be produced?

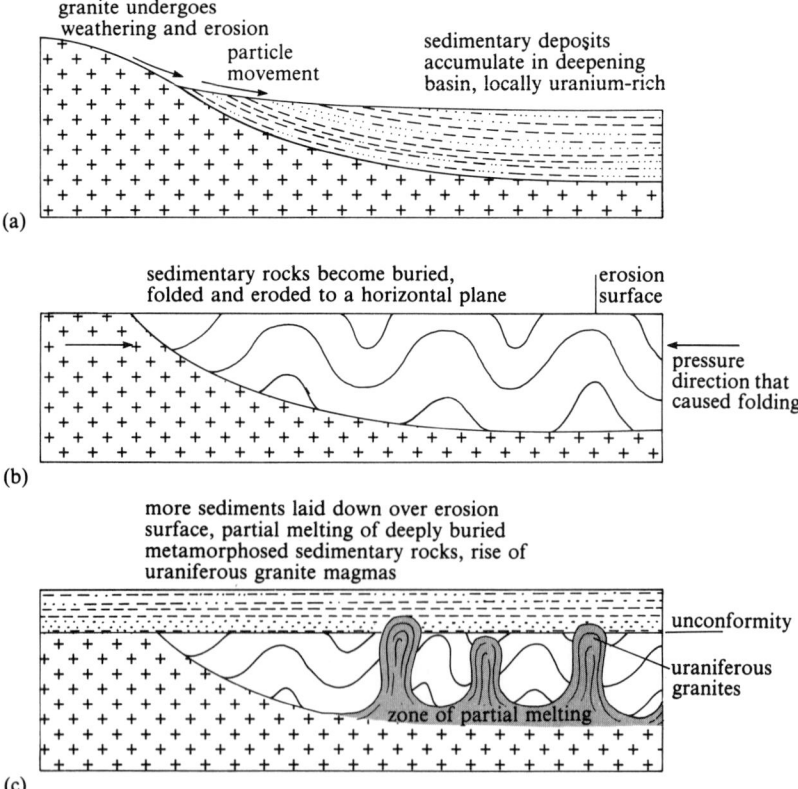

Figure 7 A schematic illustration of the sedimentary—metamorphic—igneous processes that may lead to a progressive concentration of crustal uranium to form ore deposits associated with uraniferous granites.

Image a piece of ancient *continental crust* containing granites undergoing weathering and erosion. The uranium minerals may be concentrated by surface processes (Sections 1.2.4–1.2.6), say into shales or sandstones. Now suppose that we follow these sedimentary rocks around the *rock cycle*: they may become progressively buried beneath younger sedimentary rocks for a long time — eventually they may undergo metamorphism and may reach high enough temperatures to be partially melted, forming a new generation of magmas that rise to intrude the younger formations. At this stage you should note that the last minerals to crystallize, those that contained most of the uranium in the granite that was weathered, will be the first ones to melt. So the partial melting process itself will selectively extract and enrich uranium in the magma and, of course, the new magma may concentrate uranium in its residual liquids as it crystallizes. Such geological cycles occur and may be repeated several times, resulting in rocks with economic concentrations of uranium (cf. Figure 7).

Particularly favourable regions are those underlain by *cratons* (e.g. North America, Greenland and the Baltic, also South America and Africa; Figure 8) which have been reactivated repeatedly by ancient plate tectonic processes leading to new erosion–burial–magmatism cycles as just described. The rate of crustal recycling and the surface conditions leading to uranium enrichment both seem to have been most favourable during mid-Precambrian times (see also Section 1.2.5).

The largest known deposit is at Rossing in Namibia where uraninite-bearing granites and pegmatites have been mined since 1975 in a vast open-pit operation. Similar pegmatitic uranium ores at Bancroft, Canada, occur as near-vertical veins cutting up from intrusions through Precambrian metamorphic rocks. Important disseminated magmatic deposits also occur in Greenland and Brazil, and are associated with *syenites*. Deposits in this category are generally of lower grade than other types (Table 2) but may be of large size; for example, the Rossing deposit averages only 350 p.p.m. uranium (0.035 per cent) but is

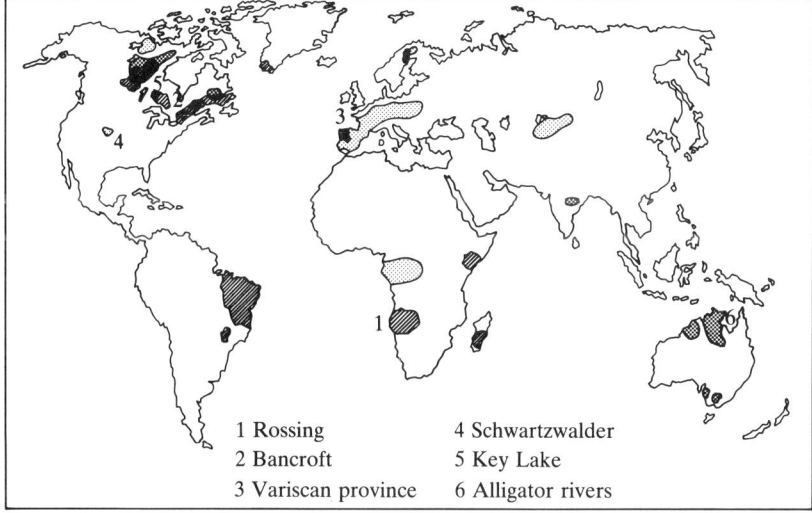

1 Rossing   4 Schwartzwalder
2 Bancroft   5 Key Lake
3 Variscan province   6 Alligator rivers

Figure 8 Location map for the world's major primary uranium deposits, showing regions characterized by disseminated magmatic deposits and pegmatites (hatched), vein-type deposits (light grey) and vein-type deposits related to Proterozoic unconformities (dark grey). Named localities are those mentioned in the text.

economic when worked on a large scale. Annual production was 3000 tonnes of uranium in the early 1980s, but the future of the mine is uncertain because of a political decision that Britain should cease importing 1000 tonnes a year of uranium from Namibia.

1.2.3 Vein-type deposits

Figure 8 also gives the distribution of two sorts of uranium deposit that occur in the form of veins: (i) those occurring in cracks, fissures, stockworks and fault zones, and (ii) a special variety of type (i) that occurs in fractures beneath Proterozoic (i.e. late Precambrian, 2500–570 Ma ago) erosional unconformities and which are considered separately because they form the largest and richest known occurrences of uranium. Together, these two sorts of vein-type deposit account for over 20 per cent of global uranium reserves and production. Deposits of the first type are of hydrothermal origin and range from massive veins to veinlets prevading zones of crushed wall rock along fault zones (*fault breccias*) several metres wide (Figure 9). These vein-type deposits are usually rich in black, massive uraninite, known as *pitchblende*, and are of medium to high grade and moderate size (Table 2).

Some of the world's earliest uranium mines were developed during the mid-nineteenth century in the vein deposits of Bohemia and Cornwall, where uranium oxides occur with tin—tungsten and sulphide mineralization. Although the Cornish deposits are no longer economic, major vein-type uranium of the same age (*Variscan,* 300 Ma) occurs throughout western Europe, notably in Czechoslovakia, France, Portugal and Spain (Figure 8). In such cases of hydrothermal vein deposits occurring in and around granites, the fracturing of the host rocks that took place during magma emplacement, or during any subsequent periods of deformation, may have provided the passageways for the mineralizing fluids rising from magmatically or other geothermally-heated sources.

Erosion surface

Figure 9 Mineralization may take place by precipitation, occasionally of uranium minerals, from solution between the crushed and angular fragments of wall rocks — the fault breccia — produced by movement along a fault zone. Arrows indicate direction of fluid flow from deep heat source.

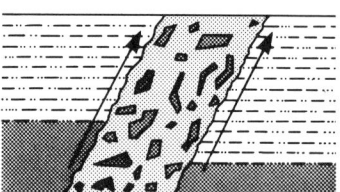

The second group of vein-type deposits are called vein–unconformity deposits, because they occur in veins beneath Proterozoic unconformities that developed during a world-wide period of mountain building about 1800–1600 Ma ago. They include major orebodies in northern Saskatchewan, Canada, and in the Alligator Rivers area of northern Australia (Figure 8). These are the largest and richest of known uranium ores and are confined to cratons and their overlying sedimentary rocks. The rocks beneath the unconformity usually include gneisses and other metamorphic rocks older than 1800 Ma, some of which are quite rich in uranium (e.g. 50 p.p.m. in granitic gneisses beneath the Proterozoic unconformity in Canada). Folding and erosion of these rocks was followed (post 1600 Ma) by unconformable deposition of late Proterozoic sandstones and impermeable shales that are still flat-lying. The latter sequence was important because it acted as a *permeability* barrier, preventing the upward migration of hydrothermal fluids from beneath the unconformity. The orebodies usually occur in rich, laterally extensive lenses (Figure 10) just beneath the unconform-

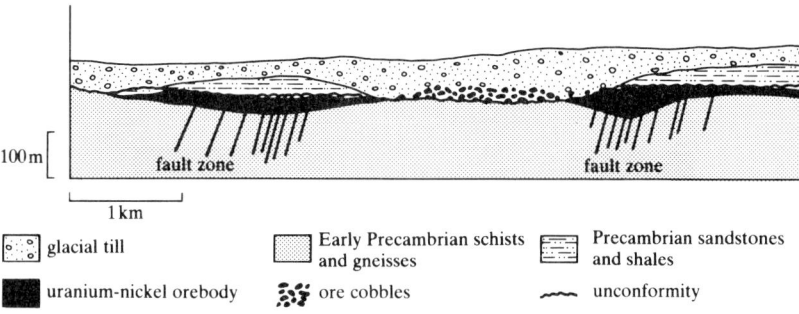

glacial till	Early Precambrian schists and gneisses	Precambrian sandstones and shales
uranium-nickel orebody	ore cobbles	unconformity

Figure 10 Diagrammatic cross-section of a vein—unconformity uranium deposit at Key Lake, Saskatchewan, Canada. Mineralization occurs where faults in the early Precambrian schists and gneisses intersect the late Precambrian unconformity which is overlain by sandstones and impermeable shales.

ity, where the metamorphic rocks are most heavily faulted and brecciated; they consist of massive and disseminated pitchblende associated with gold, nickel and other metals. The principal depositional processes responsible for ore concentration involved circulation of uraniferous hydrothermal solutions within the faulted rocks beneath the unconformity, with precipitation in *structural traps* (analogous to unconformity oil traps) (Part I, Figure 41) at the intersection between major faults and overlying impermeable, unconformable strata.

The Key Lake deposit of central Saskatchewan (Figure 10) is an example of the type of orebody resulting from these processes. It was discovered almost by accident in the mid-1970s through the recognition of airborne radiometric anomalies, due to glacially-transported boulders rich in pitchblende that occur within lakes to the south of the orebody. Subsequently, several electromagnetic anomalies over Key Lake itself, when drilled, led to the discovery of combined uranium-nickel ores beneath about 100 metres of glacial sands and gravels. Had it not been for the presence of conductive nickel sulphides and graphite schists in the fault zone, this, one of the largest known uranium orebodies, would not have been discovered. Production commenced in 1983 and is predicted to peak at 4600 tonnes of uranium a year in the late 80s/early 90s — about 12 per cent of present world uranium demand. An even more dramatic discovery of a similar orebody at Cigar Lake, further north in Saskatchewan, was announced during 1984. Already, reserves totalling 60 000 tonnes of uranium at an exceptionally high average grade of *ca*. 5 per cent have been announced. Deposits of this type, once fully exploited, will clearly revolutionize the economics of uranium production and hence of nuclear fuels (cf. Section 1.3). The obvious means of working such large, shallow orebodies is to use an open-pit method; site clearance at Key Lake involved lowering the water table by more than 100 metres, draining 15 lakes in the process, removing 100 metres of overburden, and constructing vast leak-proof tailings ponds. The return for this investment is a rich, easily worked pitchblende ore with an average grade of 1.95 per cent ranging, in places, up to 30 per cent uranium.

1.2.4 Uranium in sandstones

Although they are of low to medium grade and of modest size in comparison with many other types, uranium deposits in this category account for about 40 per cent of economic uranium resources on a world scale (Table 2). The majority occur in sedimentary rocks that were deposited in fluvial or shallow marine conditions (i.e. river systems draining onto and across shorelines), although lake deposits and continental sandstones are also occasionally mineralized. Locally, most of the uranium deposits occur as pockets, particularly in the sediments of ancient stream channels, the existence of which is deduced from distinctive cross-bedded structures confined to channel shapes containing medium to coarse-grained poorly sorted sandstones. Uranium concentrations also occur close to sandstone–shale boundaries, and particularly where abundant carbonaceous material or sulphide minerals are present.

What does the presence of cabonaceous material and sulphides indicate about the *oxidation potential (Eh)* of groundwaters within these sediments?

The oxidation potential must be negative because carbonaceous material would decay by reaction with oxygen, and insoluble sulphides would become soluble sulphates if oxidizing groundwaters had affected the sediments. As with the preservation of organic material (Part I, Section 3.1), the presence of an anaerobic, reducing environment is crucial to the formation of uranium ores in sandstone, because uranium is soluble and mobile when oxidized and is precipitated when reduced. The presence of an oxidation—reduction boundary within a permeable sedimentary sequence is crucial because it is here that secondary uranium deposits will be concentrated. Ore deposits tend to be most common where at least 20 per cent of the sedimentary sequence consists of impermeable shale. Not only do impermeable shale bands help to channel the movement of oxidizing groundwaters, but also there is usually abundant organic matter within the shale, *the oxidation and destruction of which reduces the Eh of groundwaters adjacent to the shale and causes uranium in solution to the reduced and precipitated in the adjacent sandstones.* Oxidation of sulphides in the sediments to sulphates may have the same effect. The zone of precipitation, or oxidation–reduction (*redox*) front, often takes the form of a series of C- or S-shaped ore fronts ranging from 0.5 to 10 metres in vertical cross-section through a single sandstone bed (Figure 11) — hence the term *roll-type uranium ores*. Alternatively, if there are large fragments of fossil vegetation in the sandstones, these may be completely replaced by ore as the oxidation of organic matter again provides the chemical conditions needed for the precipitation of uranium minerals. When freshly exposed, the uranium-rich rolls, or disseminated replacement pockets, consist of black minerals such as pitchblende. Subsequent oxidation on contact with air produces the spectacular yellow–orange colouration of oxidized secondary uranium minerals, many of which also contain vanadium — an element subject to similar redox — solubility controls as uranium.

Certain geological periods appear to have been more suitable than others for the formation of uranium deposits in sandstones. In particular, most deposits are confined to the post-Silurian period because land plants, which are such a necessary ingredient in providing the geochemical conditions for deposit formation, did not develop until about 400 Ma ago. Most of the known deposits (shown in Figure 12) formed around the margins of marine *sedimentary basins*, surrounded by upland areas that contained granitic rocks with above-average uranium contents. Erosion of these granites would have supplied the raw materials from which the sedimentary sequences were built (cf. Figure 7a) and subsequent groundwater migration through these sequences yielded concentrated pockets of uranium as described above.

Among the most productive deposits, those of the Colorado Plateau of North America (shown in Figure 13) contain Mesozoic and Cenozoic sedimentary sequences formed after the Rocky Mountains were uplifted, exposing for erosion uranium-rich Precambrian crystalline rocks. Some 1000 metres of shales, siltstones and sandstones were depostied on the subsiding Precambrian foundation of the Colorado 'basin', which subsequent uplift has transformed into a plateau cut by deeply entrenched river gorges. Towards the top of the sedimentary sequence occur uranium–vanadium sandstones (hence the mining town Uravan), interbedded with shales and uranium-rich volcanic rocks, which may have supplied some of the uranium that was carried into the sandstones for deposition. The sandstones are worked at several modern mines, mainly by low-cost underground drift-mining techniques, and the average grade is about 0.14 per cent uranium.

1.2.5 Quartz-pebble conglomerate deposits

Uranium deposits of this type account for just over 15 per cent of reserves and production on a world scale (Table 2). However, they are almost entirely confined

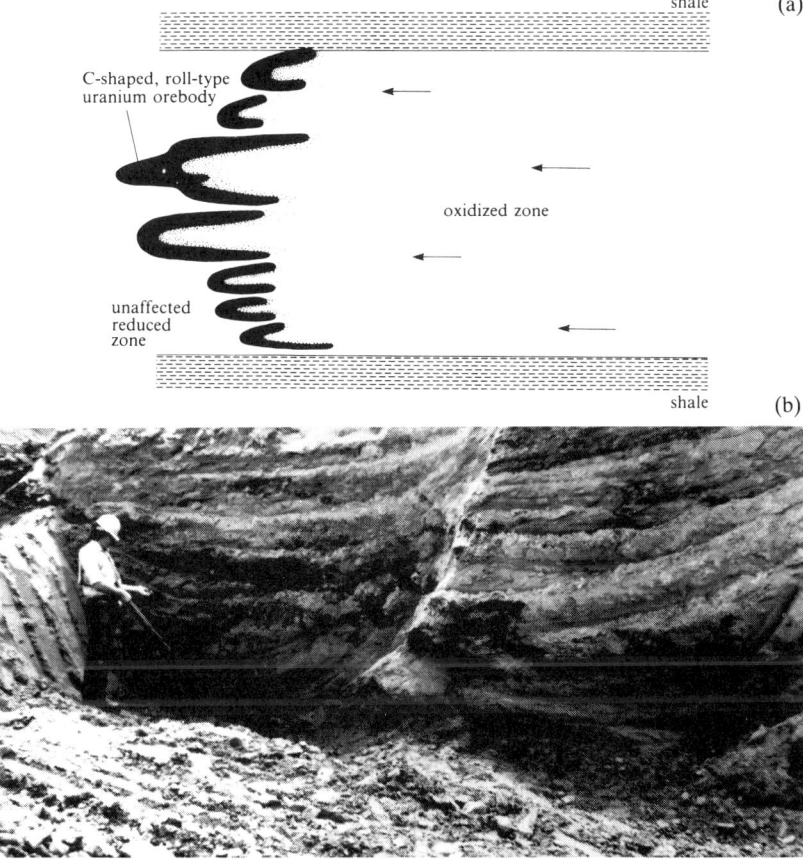

Figure 11(a)   Roll-type uranium ores produced by groundwater percolation (from right to left) in sandstone. Notice the sharp front edge of the roll produced at the oxidation — reduction boundary, and the diffuse rear edge in the oxidized zone. (b) uranium-rich roil-fronts in unconsolidated grey sands at the Pathfinder Mine, near Casper, Wyoming. This particular face shows roll-fronts pointing in opposite directions at left and right, these are connected by two horizontal zones of black uraninite. Thus, groundwaters must have percolated through a tongue-shaped zone of sandstone with uranium being mobilized towards the periphery of the tongue where conditions were less oxidizing.

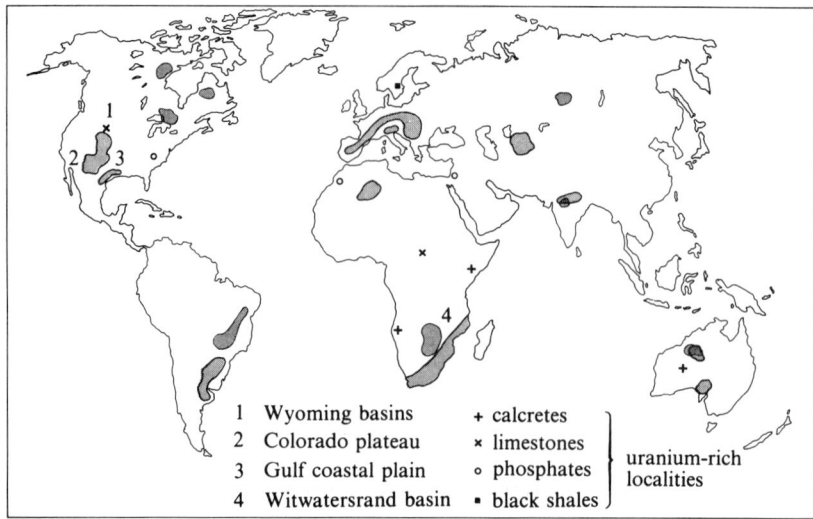

Figure 12 Location map for the world's major secondary uranium deposits showing regions characterized by deposits in sandstones (dark grey); quartz-pebble conglomerate deposits (light grey) and other secondary uranium localities mentioned in Section 1.2.6.

Figure 13 Map of the Colorado Plateau showing the location of the major sandstone uranium deposits in black.

to Proterozoic river-deposited 'placer' conglomerates overlying and derived from Archaean granitic and metamorphic basement rocks, particularly in Canada and South Africa (Figure 12). Typically the conglomerates contain pyrite, secondary micas and heavy minerals such as uraninite, or even native gold, disseminated in the matrix between the quartz pebbles, which are firmly cemented together. The rocks are medium grey in colour (Figure 14) with virtually no evidence of oxidation except by recent weathering.

This would happen if the Proterozoic atmosphere was non-oxidizing; indeed, the preservation of uraninite and pyrite in ancient conglomerates has been cited as evidence that the Earth's atmosphere was non-oxidizing before 2000 Ma ago. Other lines of evidence support this view, but during the 1980s it became more fashionable to consider the early atmosphere as *weakly* oxidizing such that these minerals were oxidized, but only slowly compared with today. The uranium grade is often so low (0.01–0.1 per cent uranium) that the element can generally

Figure 14 Quartz-pebble conglomerate sample, with uraninite in the interstices between white quartz pebbles.

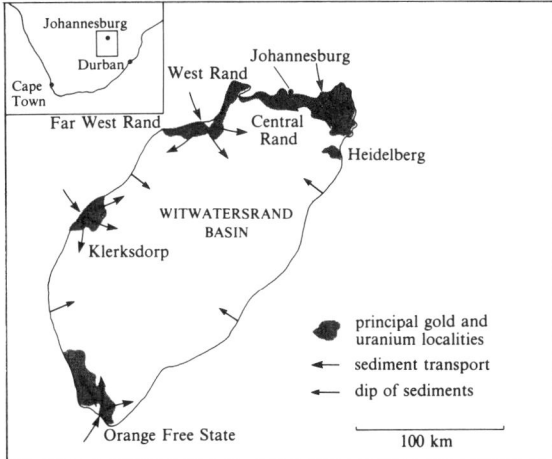

Figure 15 Distribution of Lower Proterozoic sedimentary uranium — gold placer deposits in the South African Witwatersrand basin, and their relationship to the major drainage entry to the basin.

only be worked as a by-product of some other metal, such as gold. Individual deposits are usually fairly small (5000–15000 tonnes uranium) compared with most primary and sandstone-related deposits.

A notable example of an extensive gold–uranium conglomerate deposit, some 2500–2800 Ma old, occurs in the Witwatersrand basin of South Africa (Figure 12). Figure 15 shows the directions of current flow deduced from fluvial and deltaic sediments around the margins of this Lower Proterozoic basin, which was bounded by Archaean (early Precambrian) crystalline igneous and metamorphic rocks — the source of the uraninite and gold particles. The mineralized conglomerates are located where these ancient rivers entered the basin. The deposits contain 2–5 p.p.m. gold and about 150 p.p.m. uranium; they are confined to relatively thin beds of conglomerate, known as 'reefs', ranging in thickness from a few centimetres to a metre, within a larger sequence of unmineralized sediments. In 1982, 21 mainly underground mines were operating in the area of Figure 15, producing just over 4000 tonnes of uranium metal a year, mainly as a by-product of gold mining; this represented about 70 per cent of South African uranium production.

1.2.6 Other sedimentary uranium deposits

Here we summarize uranium occurrences that are not easily classified under the ore types already considered. Although not much uranium is being produced from these sources, together they comprise about 10 per cent of economically viable uranium resources. You will find the main localities mentioned below identified in Figure 12.

Uranium in calcrete

Calcrete is a hard crust cemented by calcium carbonate ($CaCO_3$) that often forms at or near the surface of unconsolidated calcareous sediments. In hot dry climates, groundwaters carrying $CaCO_3$ in solution are drawn to the surface by *capillary action*; $CaCO_3$ is precipitated and forms a secondary calcite cement. A prerequisite for ore formation appears to be a nearby source of primary uranium, such that oxidized uranium in solution may be carried in the same groundwaters that form the calcrete. Precipitation of oxidized secondary uranium minerals occurs where the *porosity* of the sedimentary formations has been decreased through calcrete precipitate, so that the flow rate of uranium-bearing groundwaters is also decreased. Calcrete-associated deposits have been discovered in Australia, Namibia and Somalia — they have yet to be fully assessed but are potentially important; for example, the deposits at Yerilla, Western Australia, are estimated to contain 40000 tonnes of uranium at a grade of 0.12 per cent uranium.

Uranium in limestone caverns

Small deposits (Table 2) of secondary uranium minerals precipitated from groundwater may form linings to passageways and fractures in limestones. For example, the Carboniferous limestones of Wyoming are characterized by a complex cavern system, filled by carbonate-rich silts that contain uranium minerals.

Uranium in phosphate

These deposits may be either (i) accumulations of fossil bones in marine or continental strata, or (ii) chemically precipitated phosphatic nodules in marine sediments. Reducing conditions, producing the U^{4+} ion in uraniferous surface waters, are important because U^{4+} can replace Ca^{2+} in the principal mineral, apatite, $Ca_5(PO_4)_3(OH,F)$. The charge difference of 2+ is compensated by adjustments in the proportion of OH^- and F^-, which are both rather loosely held in the apatite structure. The concentration of uranium in phosphates is usually very low, typically 100–150 p.p.m., and so it is not usually economic to mine such deposits for uranium alone. But phosphates are widely worked for other purposes in north Africa, the Middle East and the USA: uranium is extracted as a by-product.

Uranium in coal and black shales

The reducing environments in which coal and black shales form (lagoons and coastal swamps) will also be ideal for the precipitation of uranium-rich compounds. Figure 16 shows some uranium concentrations in Coal Measure deposits: not surprisingly, the most reduced members of the sequence contain the most uranium. Most deposits are poor, but some lignites in North Dakota and coals from Europe contain between 500 and 2000 p.p.m. uranium. Severe difficulties in extracting uranium from these coals and lignites make them uneconomic for mining, but viable uranium deposits in this category do occur in bituminous shales, particularly in the Ranstad area of Sweden. Although the ore grade is only 300 p.p.m., the extent of this deposit, over 500 km^2, makes it one of the world's richest deposits, with other 500 000 tonnes of uranium. Nevertheless, at present, no production is allowed because of a veto by local authorities for environmental reasons.

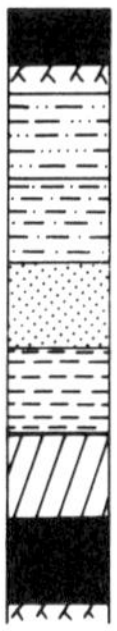

LITHOLOGY	ENVIRONMENT	CHEMICAL CONDITIONS	URANIUM CONCENTRATION/p.p.m.
coal	swamp	reducing	20–6000
seatearth (fossil soil)			0.5
sandy shale			2–4
siltstone	delta	oxidizing	1.5
sandstone			1.5
black shale and sulphides	lagoon	reducing	4–40
black calcareous shale with sulphides	marine		20–1200
coal and sulphides	swamp	reducing	20–6000

Figure 16 Idealized sediment cycle from a Coal Measures sequence (cf. Part I, Figure 7), with details of the environment of deposition and the ranges of likely uranium contents (in parts per million of uranium).

Before we conclude this survey of uranium deposits, you should realize that not all sedimentary formations that accumulate under reducing conditions are uranium-rich. No matter how ideal the conditions may be for the precipitation of uranium minerals, they cannot form unless there is dissolved uranium in waters flowing into the depositional basin.

Figure 17 gives a simplified summary of the relationship between the main categories of uranium deposits, showing how they are associated with different parts of the rock cycle and illustrating the geological processes that are responsible for producing them. You should study Figure 17 in conjunction with Table 2, which gives the grades, sizes and proportion of world-wide reserves for each category of deposit.

1.3   Uranium production and economics

Study Comment. So far we have considered uranium as an energy resource and have examined the wide variety of geological processes that produce mineable uranium ore deposits. This Section follows ore from the mine to the mill, where the product is in a suitable state for use in power stations, and then considers trends in uranium price, production and reserves.

1.3.1   Mining and milling of uanium

The geological diversity of uranium deposits is reflected in mining techniques: *stope* mining for veins, *room and pillar* mining for buried stratiform deposits and open-pit mining for other stratiform deposits and some disseminated magmatic and vein—unconformity deposits. Of course, the mining technique

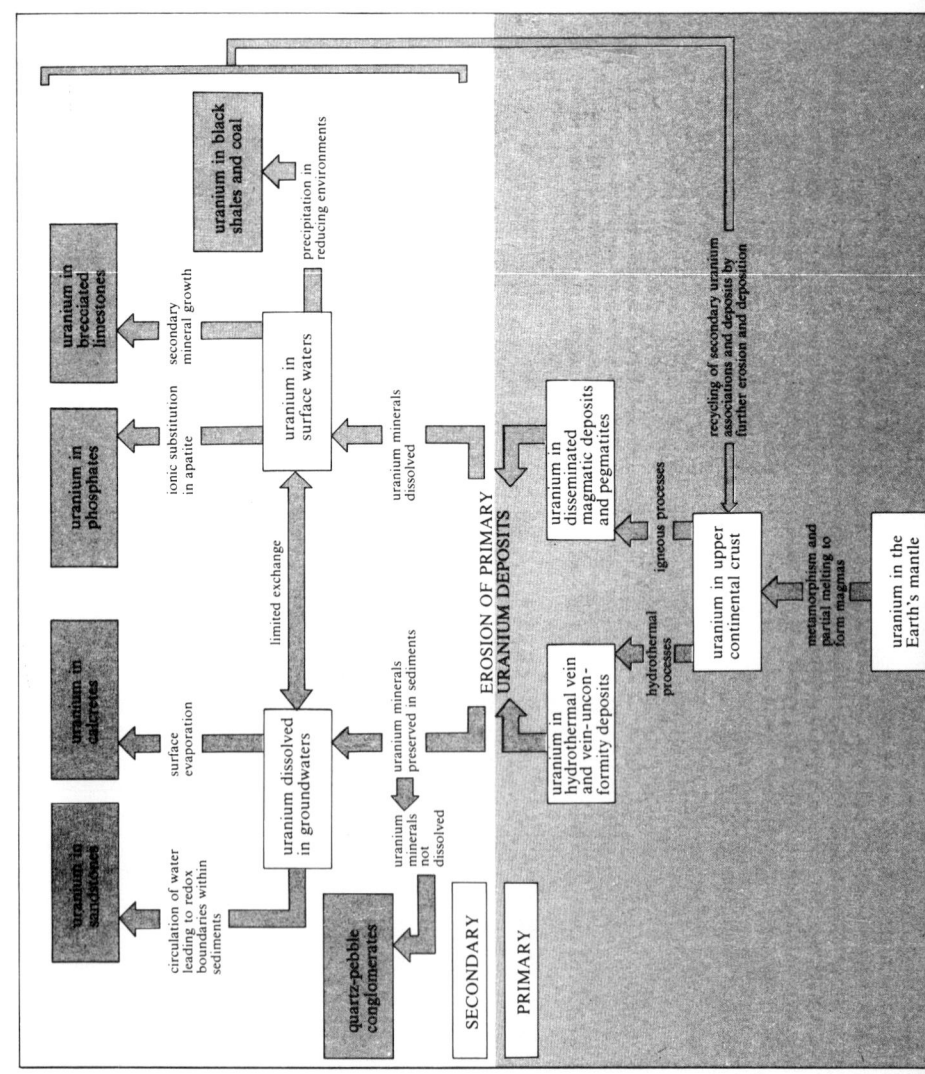

Figure 17 Conceptual diagram of the uranium cycle, showing the sources of uranium and the causes of its mobilization and stabilization into the six main classes of ore deposit discussed in Section 1.2.

has an important effect on the economic viability of a deposit. The operating costs of underground mines are higher for every tonne of ore produced than are those of open-pit mines. But there are much smaller differences between the two in terms of capital costs, haulage costs, taxes and royalties. A comparison of typical ranges of costs under these headings for underground and open-pit mines in the USA is given in Figure 18a.

Once the ore has been mined, it is transported to a mill where it is crushed and the uranium is leached into solution, separated in ion-exchange columns, precipitated and dried to form uranium oxide for shipment, known as *yellowcake* (see Figure 19 for details). The cost of concentrating uranium in a mill adds between $12 and $40 per tonne of ore to production expenses (Figure 18b) and is, of course, independent of the type of mining operation. The *combined* costs of mining and milling (Figure 18c) are, on average, $51 per tonne of ore from an open-pit mine and $111 per tonne of ore from an underground mine. As you can see from Figure 18a, the main reason for the difference is that the cost per tonne of operating an underground mine is three

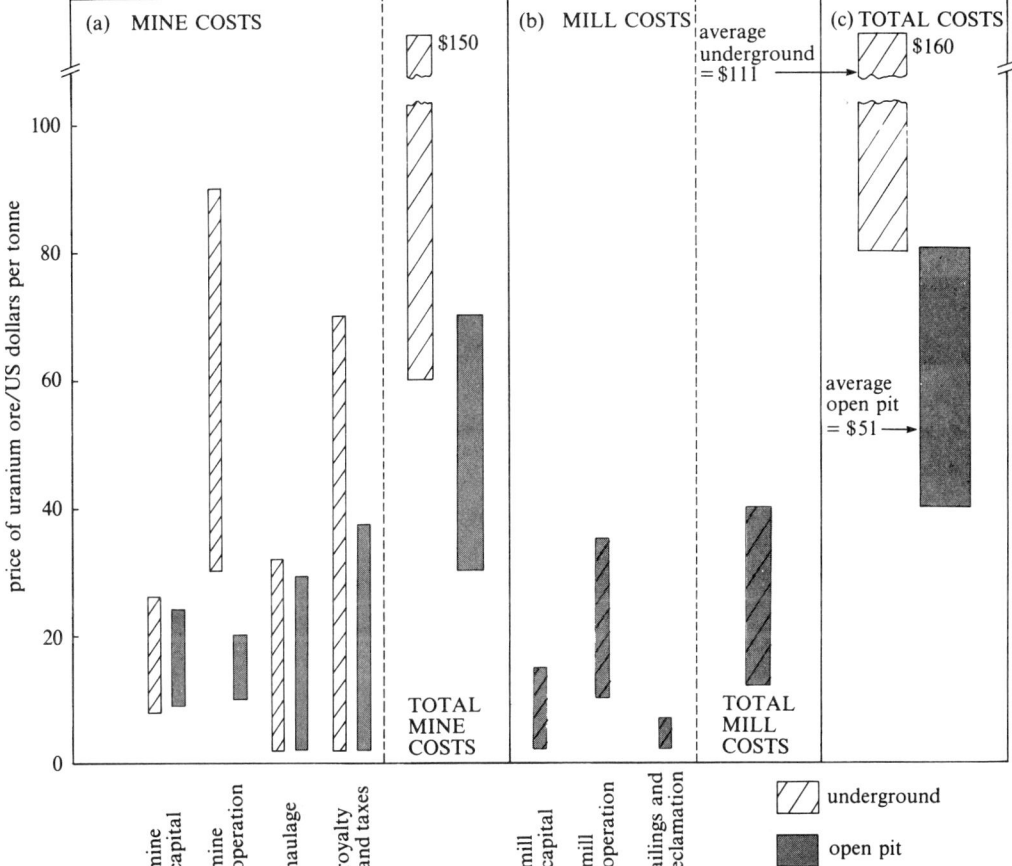

Figure 18 Comparison of the ranges of costs (1981 prices) for uranium production by underground and open pit operations in the USA. Note that costs are given in dollars per tonne of ore processed — the higher the ore grade, the lower the cost per tonne of uranium produced. The mill costs are the same for both types of operation, so it is the difference in mine costs that makes underground methods more expensive. Although most costs have increased since 1981, the principles reflected in this illustration are unchanged.

or four times greater. With this kind of cost differential you might wonder why, with plenty of ore available, there are any underground uranium mines operating today. The answer is 'ore grade' (remember that the costs shown in Figure 18 are per tonne of *ore*) – it is perfectly reasonable to mine underground if the grade of ore is high enough to offset the mining cost disadvantage.

This concept can be illustrated by calculating the minimum grade of ore which would be required to 'break even' in open-pit and underground mining operations. We will assume a uranium price of $50 kg^{-1} (average for 1982–1985) and, for simplicity, 100 per cent recovery. In the average open-pit mine it costs $51 per *tonne* to produce uranium which is sold at $50 per *kilogram*. Therefore, just over 1 kg of uranium must be produced from every tonne of ore mined. This is a grade of just over 0.1 per cent uranium (1000 p.p.m.). In contrast, the average underground mine produces a tonne of ore for $111 and so we need to generate 111/50 = 2.22 kg of uranium from each tonne to break even. This is a grade of 0.22 per cent uranium (2200 p.p.m.). Of course, both grades are minimum estimates because the running costs given in Figure 18 are continuously increasing, because recovery will be much less than 100 per cent and because post-milling transportation costs are not negligible. Nevertheless, the fact remains that much higher grades of ore are required for economic underground mining compared with open-pit methods. Figure 20 shows that the market price of uranium peaked at $110 kg^{-1} in 1979. At that time, uranium mining was a boom industry undergoing rapid expansion with the opening of new mines and a doubling of world production within the preceding 5 years. During this period there was an eight-fold increase in the market price of uranium.

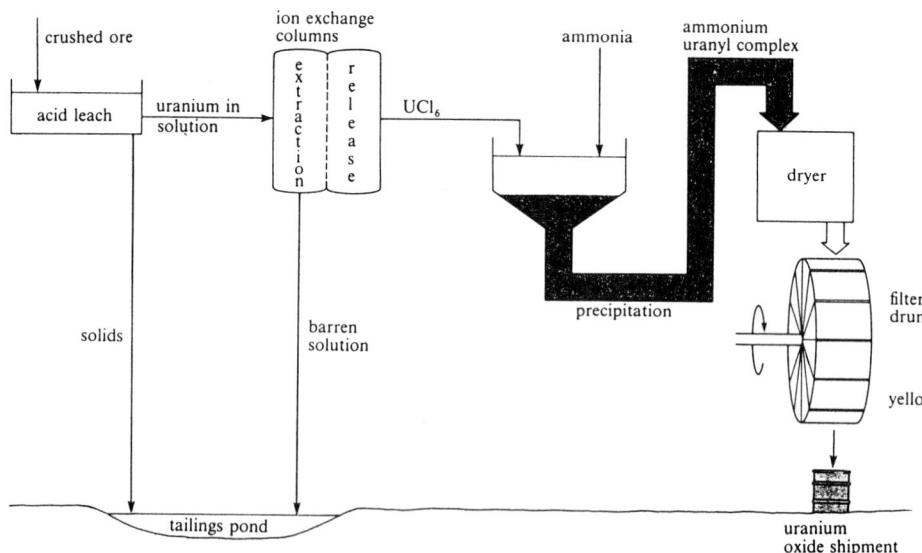

Figure 19 Simplified flow diagram showing the operation of a uranium mill. Uranium minerals are leached from the crushed ore in an acid bath and are then extracted from other acid-soluble material in ion exchange columns. Uranium is released as uranium chloride and then precipitated as an ammonium salt which is dried and roasted to produce 'yellowcake' oxide for shipment.

Can you identify the principal reason for this upward trend in uranium price and production?

The 'oil crisis' of 1974 and subsequent years focused attention on the urgent need to develop non-hydrocarbon energy resources (cf. Part I, Section 2.2) Uranium energy resources were an obvious potential growth area and, accordingly, nuclear power stations were planned and many were developed throughout the western world (cf. Section 1.1). Deposits that were previously uneconomic suddenly became viable, and widespread exploration led to the discovery of new deposits on all continents (Section 1.3.2). For example, the Rossing deposit in Namibia (Section 1.2.2), working the lowest known uranium grade to have proved viable with the least expensive, large scale open-pit methods, started production in 1975. Over a period of 5–6 years the uranium-producing industry responded to the relatively rapid increase in price and demand during 1974 and 1975.

Why then was there an about-turn in the fortunes of the industry during 1980 and subsequent years?

The anticipated growth in energy demand during the late 1970s just did not happen and many of the plans for new nuclear power stations were cancelled. There were also important environmental objections to the rapid development of nuclear power (cf. Section 4). These factors combined to result in over-production of uranium, leading to stockpiles and a consequent slump in the

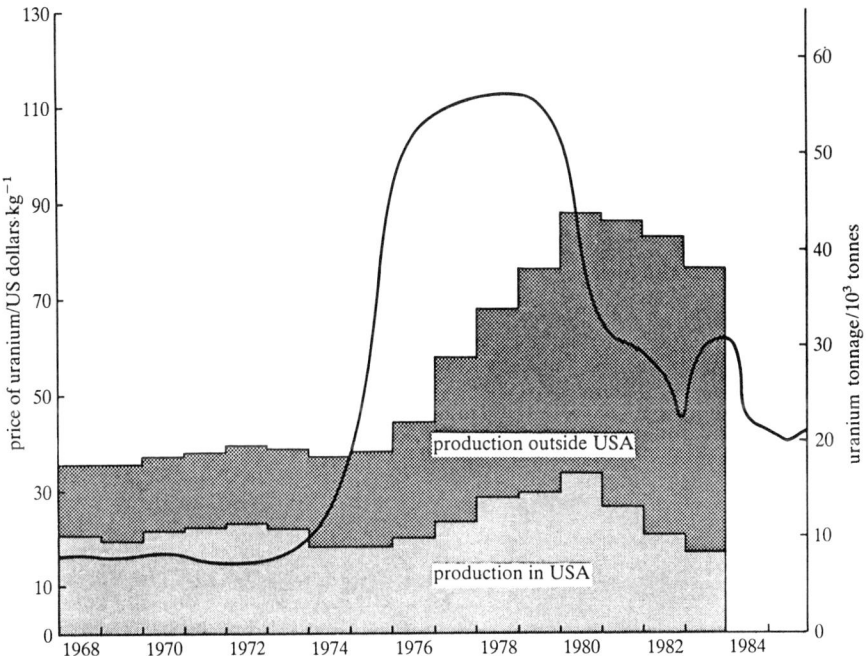

Figure 20 Uranium prices and production statistics from 1968 to 1983. The left-hand axis and black curve refer to trends in the market price of uranium. The histograms for production in the USA and the rest of the world relate to the right-hand axis; for example, total world production in 1980 was almost 45 000 tonnes.

industry, characterized by a halving of the uranium market price in 18 months during 1979–80 before some stability returned in 1982–85. However, the market price of uranium has remained volatile and the 1984 slump, in particular, reflects the input of cheaply produced uranium from vein-unconformity deposits onto the market. Since peaking in 1980 the production histogram (Figure 20) has responded surely but slowly, and during 1981–85 many mines became unprofitable and were closed. Those with marginally economic and sub-economic grades that were able to survive the slump did so because they were large, or because they were regarded as being of strategic national importance, or because they had contracted for long-term fixed price sales during the boom period.

1.3.2 Uranium reserves and resources

In the uranium industry, uranium *reserves* and *conditional resources* are usually referred to as *reasonably assured resources*. These include uranium that almost certainly exists and can be mined at a price which may (reserves) or may not (conditional resources) be economic, depending on the state of the market. For example, five categories of reasonably assured resources are shown for the USA in Figure 21, those classified at less than $20 kg^{-1} being the most economic to mine and those at $130–260 kg^{-1} being the least economic. Study Figure 21 carefully and notice the following three points.

1   The graph is a histogram which shows, *for each year*, the tonnage of uranium resources that fell into each of the price categories at the beginning of the year. To make sure that you can use the information in Figure 21, estimate the resources in each of the three categories represented on the graph for the year 1982.

The $40–80 kg^{-1} category contained 360 000 tonnes, the $80–130 kg^{-1} category 250 000 tonnes and the $130–260 kg^{-1} category 360 000 tonnes.

2   You should now appreciate that the estimates of resources increased progressively throughout the period of intensive exploration and increased production during the late 1970s, reaching a peak in 1980.

3   The amounts of low-cost uranium resources have decreased progressively since 1972–74 because of increased mining costs which have shifted resources in the less than $20 kg^{-1} category into the $20–40 kg^{-1} category, and so on. By 1983 there were no worthwhile US uranium resources left in the categories less than $40 kg^{-1} and those in the $80 kg^{-1} were shrinking rapidly such that, with a 1985 price only a little above $40 kg^{-1} (Figure 20), there are almost no economic resources left in this, traditionally the world's largest uranium producing country. This is reflected in the rapidly declining US production data for the post-1980 period (Figure 20).

However, it is common practice among uranium-producing nations to quote resources in the same price categories as used for the USA in Figure 21. Estimates of resources in the less than $80 kg^{-1} and $80–130 kg^{-1} ranges appear in Table 3 for the eight non-communist nations that are likely to be the main uranium producers for the remainder of this century. Resources priced at more than $130 kg^{-1} are not included in present estimates for the obvious reason that they are unlikely to become economic reserves in the foreseeable future (cf. uranium price trends in Figure 20).

Table 3 also includes data for a second major class of uranium resources — some of those in the *hypothetical resources* category which, by tradition, in the

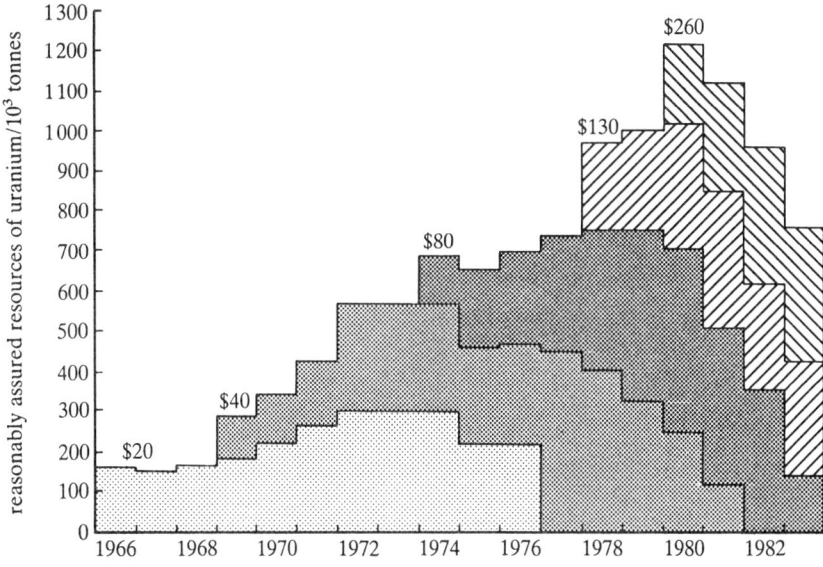

Figure 21 Historical changes in the reasonably assured resources of uranium metal in the USA by category of estimated extraction cost (dollars per kilogram) for each year from 1966 to 1984. As production costs have risen, resources from lower cost categories have moved into higher cost categories. (So the disappearance in 1977 of uranium priced at $20 kg^{-1} does not necessarily imply that *all* the resources in this category during 1972–1974 had been extracted, merely that some had changed to higher cost categories.)

uranium industry, are known as *estimated additional resources*. These are the poorly known resources that may form extensions to well-explored deposits and also include those that may just not have been measured and assessed because of inaccessibility and/or lack of strategic national importance. These estimated additional resources have been assigned to the same two categories: less than $80 kg^{-1} and $80–130 kg^{-1}. A glance at the totals for each category shows that the inclusion of estimated additional resources could substantially increase the available uranium. For the moment, however, we shall concentrate on the distribution and use of reasonably assured resources.

The pre-1984 statistics for production and reasonably assured resources in Table 3 are shown in graphical form in Figure 22 fo easy reference. Look first at the geographical distribution, and notice that 75 per cent of uranium resources occur in the continents of North America, Africa and Australia. In view of the information in Figures 8 and 12 this probably caused you no great surprise.

But why do just a few countries dominate the resources picture? Can you think of two reasons?

The first reason is geological. You may recall from Section 1.2 that regions such as the three continents identified above, which are underlain by Precambrian cratons, are often the most fruitful areas for uranium mineralization. This is because their long geological histories have provided ample opportunity for repeated concentration, particularly during Proterozoic times when conditions for uranium concentration were ideal. On this basis there should be no geological reason why other cratonic areas for example, in Asia, South America, Greenland and Scandinavia, should not be equally endowed with uranium deposits.

Table 3 Uranium resources and pre-1984 cumulative production from principal uranium-producing countries (excluding China, USSR and associated countries). All data are in thousands of tonnes of uranium metal

Country	Pre-1984 production	Reasonably assured resources		Estimated additional resources	
		less than $80 kg^{-1}	$80–130 kg^{-1}	less than $80 kg^{-1}	$80–130 kg^{-1}
Brazil	1	163	0	92	0
USA	300	131	276	30	52
Canada	162	176	9	181	48
France	56	75	16	28	15
S Africa	111	191	122	99	48
Namibia	25	119	16	30	23
Niger	30	160	0	53	0
Australia	22	314	22	369	25
others	33	139	114	32	97
totals	740	1468	575	914	308

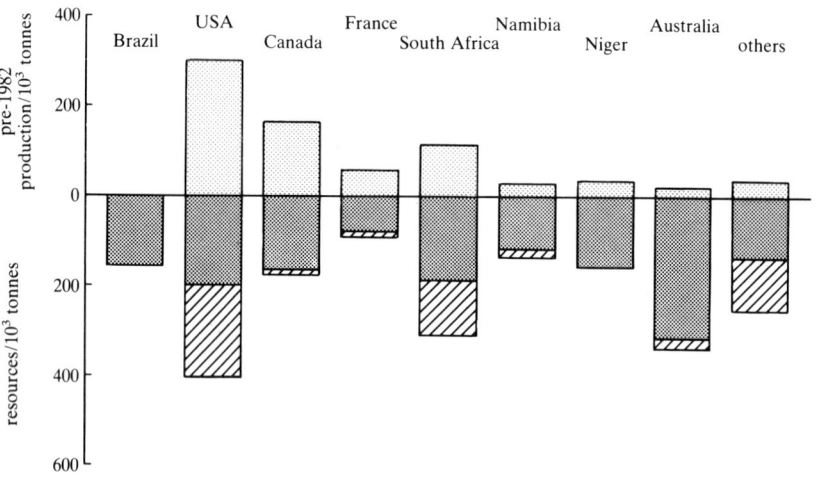

Figure 22 Comparison of pre-1982 production (fine stipple) from the world's major uranium-producing countries with their reasonably assured resources in the categories of less than $80 kg^{-1} (heavy stipple) and $80 — 130$ kg^{-1} (ruled area).

This brings us to the second possible reason for the predominance of North America, Africa and Australia — these are the areas that have seen the most intensive efforts to discover and produce their resources. If the criteria used to assess known uranium deposit were to be applied throughout the non-communist world, then the uranium resources available at less than $130 kg^{-1} might be increased by some 7 to 15 million tonnes. These speculative figures were produced by the International Uranium Resource Project in 1980, and by the same criteria, eastern Europe, the USSR and China should be able to add a further 3–7 million tonnes of uranium. Clearly, this would greatly increase the amount of uranium available (only about 3 million tonnes are quoted in all categories of Table 3), and thence the lifetime of the total resource bank. However, as prices continue to increase it would be more prudent to regard world hypothetical resources at a (1985) price of $130 kg^{-1} or less as between 5 and 10 million tonnes.

Now consider the other set of data in Table 3. What is the main factor responsible for differences in pre-1984 production between the various nations?

Again, the level of historical production is connected to the amount of past exploration and must reflect the length of time during which each nation has been interested in producing uranium. Although the USA still has a larger proportion of the world's resources than any other country (Table 3 and Figure 22), it has been producing uranium for several decades, together with Canada, France and South Africa. By 1984 these four countries had each produced between 20 and 50 per cent of their total uranium in the class of reasonably assured resources. Namibia, Niger and several 'other' countries which had produced only about 10 per cent of their resources in this category by 1984, did not become uranium-producing nations until the major growth period during the 1970s. The contribution of these countries largely accounts for the increasing proportion of uranium production from outside the USA during the period 1975–80 illustrated in Figure 20.

Finally we focus on Brazil and Australia, which produced relatively little uranium before 1984. Large-scale exploration in Brazil commenced in 1974 and led to the discovery of several big deposits, the first of which came into production at a low level during 1981. Australia has a much older uranium industry: exploration started in 1947 and, although several orebodies were discovered, Australia did not join the league of large-scale uranium producers until the end of a second phase of more intensive exploration during the 1970s. Like Canada, Australia has several very rich, easily accessible uranium deposits that are cheap to work and which, therefore, may have a major impact on world markets during the next 10–20 years. Most of this impact is likely to be sustained in the USA, where present levels of production could be maintained only if severe import controls were to be imposed. Some experts who favour such controls argue that the thousands of uranium workers who were laid off during the early 1980s (owing to mine closures — cf. Section 1.3.1) will be required when the 81 new reactors in the USA in various stages of construction and licensing are added to the 78 that were operating in 1983.

Others maintain that in order for the electricity companies to be profitable, it would be better to import cheap uranium than to produce more expensive uranium at home. A third group foresees that many of the new reactors will not be commissioned because of public concern over health and safety, following incidents such as the leak of radioactive materials from an overheated reactor core at Three Mile Island (Pennsylvania) in 1979 (cf. Section 1.1.1.) and at Chernobyl (USSR) in 1986.

What, then, does the future hold for the uranium industry? At best, the future is highly uncertain because every nation that produces and/or consumes nuclear fuel needs to make balanced judgements about the kinds of argument outlined above. The increasing use of uranium for high density warheads may affect these arguments in the near future — although depleted uranium that has been through a burner reactor can be, and often is used for this purpose. However, there is no doubt that mid-1970s projections of the number of nuclear power stations and the associated uranium demand were grossly overestimated. With the proviso that there may already have been considerable changes as you read this text, Figure 23 gives a compilation of planned production capabilities for the period 1984 to 1995. If you compare the real (i.e. historical) production data in Figure 20 with these projections, you will see that planned production in 1984 (about 45 000 tonnes) exceeds actual production in 1983 (about 38 000 tonnes) by a large margin, suggesting that the forecasts of production later in the 1980s in Figure 23 may also be overestimated.

Earlier, in Section 1.1 and Figure 4, it was shown that an average 1000 MW burner reactor running for 30 years requires 4800 tonnes of uranium fuel, or an average of 160 tonnes per year. The data in Figure 23 indicate that, also on average, some 50,000 tonnes of uranium per year could be produced during the early 1990s. This means that $50 \times 10^3/160 = 312$ reactors could be fuelled from produceable uranium. Now the power output of those reactors will be 312×1000 MW, or 312 GW, which is equivalent to $312 \times 10^9 \times 3.15 \times 10^7$ joules per year, i.e. just short of 10^{19} J yr^{-1}.

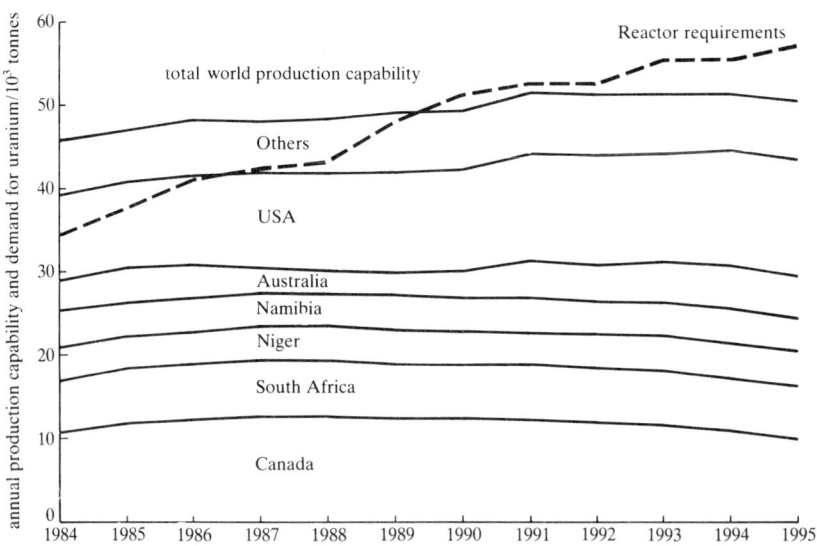

Figure 23 Estimates of annual reactor uranium requirements (dashed curve) superimposed on projected uranium production capabilities (solid curves) using known resources recoverable at a cost of $130 kg^{-1} uranium or less, for the period 1984 to 1995.

There are several important points, arising from this calculation. First, if you consult the world energy production forecasts given in Part 1, Figure 4, you will find that some 2×10^{19} J yr^{-1} would, according to that scenario for energy demand, be required in the early 1990s, nearly twice the power production that can be sustained from the available uranium. A second point is that there are also insufficient reactors built or in construction, to achieve a much greater nuclear production than 10^{19} J yr^{-1} by the early 1990s. The OECD (Organisation for Economic Co-operation and Development) Nuclear Energy Agency recently carried out an assessment of world reactor uranium requirements (1985–1995) and their best estimate is reproduced as a dashed curve superimposed on Figure 23. This curve clearly passes through the planned uranium production ceiling just before 1990, implying either that more uranium will have to be produced, or that there will be insufficient uranium to fuel all the available reactors. This is in total contrast to the over production and stockpiling of uranium that occurred during the early 1980s and which led to the price slumps of 1982 and 1984 (Figure 20). On the basis of the information presented so far we can deduce that, barring further reaction to the anti-nuclear lobby as after 1979, the overall fortunes of the uranium mining industry are likely to improve during the 1985–95 period. Remember, however, that the production data given in Figure 23 are based on a uranium price of $130 kg^{-1} whereas the 1985 price was ca. $40 kg^{-1}.

None of the USA contribution to Figure 23 could be produced economically at the 1985 price. Some relative shifts in the national contributions to world production may therefore be anticipated and, in particular, the full impact of vein-unconformity deposits on Canadian and Australian production is not reflected in the most recent projections incorporated into Figure 23. Another factor is that continued mine closures during the 1984–88 period, while the market price may remain low (because production potential outstrips demand), will also have a major impact on relative national contributions. At best, all we conclude is that the uranium supply and demand picture is likely to remain rather volatile for the foreseeable future, and this is hardly a basis on which to build reliable *long-term* forecasts of world nuclear energy production beyond the early 1990s.

In general, what will happen beyond that time depends on whether growth in demand for energy is maintained, in which case more nuclear plants may be required, and on the potential for widespread development of commercial fast breeder reactors. Figure 24 summarizes various predictions of the growth in demand for uranium for the period up to 2025. The heavy broken and solid curves, both derived from the OECD multi-nation survey, represent the data for existing centres (Figure 23) and maximum attainable uranium production. The fine solid and broken curves are OECD high and low growth cases extended to 2025. These curves split to show alternative uranium requirements(a) if only burner reactors are used and (b) if a mixed breeder–burner strategy is adopted from about AD 2000 whereby burners are used to provide plutonium and depleted ^{238}U for breeders. Assuming that some of the planned and prospective production centres are introduced, there is enough uranium for these curves to be followed until the year 2000. However, the OECD high growth case could only be realized in the period 2000–2025 if new resources are added, whereas the low growth case could survive if, simply, breeders are introduced and today's estimated additional resources become 'assured'. On the timescale in question there is no reason why this should not happen and, in that case, some of today's hypothetical resources would, through further exploration, have entered the assured or additional resource categories.

Finally, what are the implications of the two OECD nuclear growth cases for power supply in the year 2010 when, according to Figure 4 in Part I, about 6×10^{19} J yr^{-1} should be produced from nuclear sources? The two sets of curves pass through 1000 and 625 GW for high and low growth — equivalent to 3.2 and 2.0×10^{19} J yr^{-1}, respectively. As with the data for 1990 (see earlier

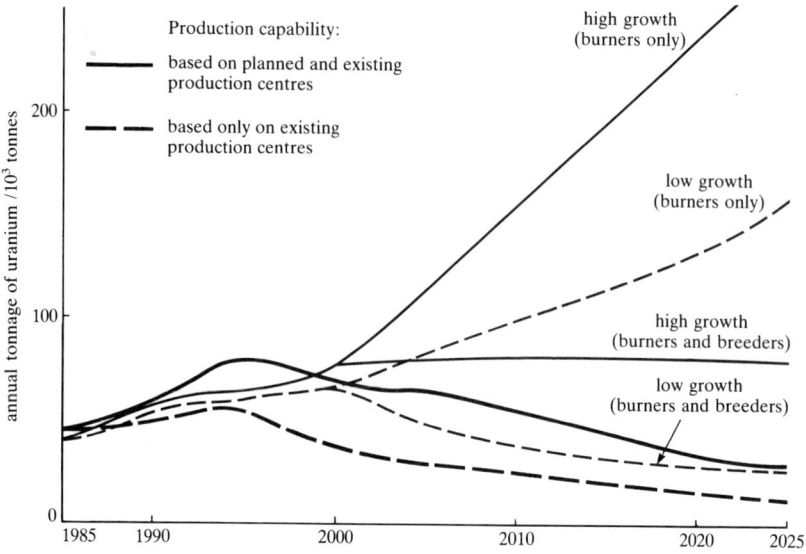

Figure 24 Estimates of the world supply (heavy curves) and demand (fine curves) for uranium between 1980 and 2025. The area under the broken heavy curve represents currently available resources (1985) in both 'reasonably assured' and 'estimated additional' categories at less than \$130 kg^{-1} which can be produced from existing and committed (i.e. under immediate development) production centres and is identical to the uppermost curve in Figure 23. The heavy solid curve gives an estimate of world uranium production that could be achieved if planned (recoverable resources) and prospective (estimated additional resources) extraction centres are also introduced. Two pairs of demand curves are shown, first for high growth on OECD projections (fine solid lines — producing *ca.* 1000 GW by AD 2010) and second for a low growth case (fine broken lines — producing *ca.* 625 GW by AD 2010). Note that these long-term projections are highly speculative and should not necessarily be regarded as the extremes of a spectrum. Note also that the reference strategies (i.e. burners only, or burners plus breeders) are intended to illustrate the extreme range of uranium requirements that might result if a particular growth scenario in nuclear power occurs.

calculation) even the high growth case, which looks extremely overstated at the time of writing, does not quite reach the projected demand. Although these comparisons are based on speculative projections and should be treated with caution, it is clear that a rapidly developing nuclear industry would be required if the 'energy gap' referred to in Section 2.2 of Part I (a) turns out to be real and (b) is to be filled by a major nuclear contribution.

Section 4.2 discusses some of the environmental reasons, connected with waste storage, nuclear accidents, etc., why governments may find such strongly pro-nuclear decisions difficult to make.

1.4 Summary of Chapter 1

1   Nuclear power generation in reactors results from the accelerated fission of heavy uranium isotopes when bombarded by neutrons. Conventional burner reactors, such as Magnox, AGRs and PWRs, require the relatively scarce ^{235}U isotope whereas fast breeder reactors, which are still being developed, make much more effective use of available uranium fuels by exploiting ^{238}U.

2   A 1000 MW burner reactor 'burns' 4800 tonnes of natural uranium fuel over a lifetime of 30 years whereas the equivalent breeder reactor requires about 53 tonnes of uranium and plutonium fuel.

3   The principal geochemical characteristics that determine the distribution and concentration of uranium to form ore deposits are: (i) the large charge and size of uranium ions, which cause them to accumulate in the residual magmas and watery silicate fluids associated with magmatic processes; (ii) the much greater solubility of U^{6+} than of U^{4+} in surface environments, which promotes uranium transport in oxidizing groundwaters and deposition from reducing groundwaters.

4   Although many uranium deposits result from repeated cycles of concentration through various crustal processes, those last associated with internal processes (e.g. dispersed magmatic and hydrothermal vein deposits) are termed 'primary', whereas those last associated with surface processes (e.g. sandstone and conglomerate deposits) are termed 'secondary'.

5   The relative importance, in terms of reasonably assured reserves, grade and size, of the six main categories of uranium ore deposits is summarized in Table 2 and Figure 17.

6   The relatively high cost of underground compared with open-pit mining (Figure 18) can be offset by differences in ore grade. Recently discovered vein–unconformity deposits of high grade, which are amenable to open-pit mining, could dominate the uranium markets for the foreseeable future.

7   Reasonably assured and estimated additional resources of uranium which total, respectively, about 2.0 and 1.2 million tonnes (Table 3) are located in well-explored countries underlain by Precambrian cratons. Hypothetical resources could increase the total by 5–10 million tonnes, all of which might be produced at less than $130 kg^{-1} (1985 prices).

8   Uranium exploration and production during the 1970s and early 1980s anticipated a much greater demand than was forthcoming, a situation which has led to uranium stockpiles, a depressed industry and the closure of mines with low grades but no fixed price contracts. Present assured and estimated resources are therefore adequate to fulfil the forecast demand until well into the period 1990–2000 (Figures 23 and 24).

9   OECD projections of the contribution of nuclear energy lie in the range 2.0–3.2 $\times$ 10^{19} J yr^{-1} by 2010, and if the upper end of this range is to be achieved, hypothetical uranium resources must become assured and recoverable, and a mixed burner-breeder reactor strategy is almost essential.

2 Geothermal energy

Study Comment. In this Chapter we deal with techniques of energy production that depend entirely on heat produced *within* the Earth, mainly by the decay of long-lived radioactive isotopes of uranium, thorium and potassium. In that sense, like uranium resources for the nuclear power industry, we are dealing here with another *primary, internal* source of energy.

The picture that the term 'geothermal energy' may bring to mind is one of boiling mud pools, geysers and vents in volcanic areas gushing with hot steam. Traditionally, most attempts to exploit geothermal energy have relied on the drilling of boreholes in such areas to tap steam, at high pressure, which is then used to drive turbines that generate electricity, in the same way that steam raised by coal combustion is used in conventional power stations. Alternatively, the steam may be piped throughout the surrounding area for direct use in space heating and industrial processes. During the 1970s, however, two new techniques were developed for exploiting geothermal energy in areas with less obvious geothermal resources. The first of these exploits warm water pumped from well-insulated deep *aquifers*, again for domestic and industrial heating; the second seeks to produce steam for electricity generation by drilling more deeply than usual to encounter suitable hot rocks.

The flow of internally generated heat across the Earth's surface is fairly constant over most of the globe, averaging about 60 mW m^{-2} (milliwatts per square metre). But in certain places, where there is a close association between hot springs and volcanoes, the flow of heat may be much greater, up to 300 mW m^{-2}.

Where, in terms of their distribution within the tectonic plates, are such thermally anomalous regions likely to be concentrated?

Most of the world's volcanoes occur at the boundaries of the tectonic plates, either at *constructive plate margins*, where new ocean lithosphere is formed as basaltic magma rises to the surface (e.g. in Iceland), or at *destructive plate margins*, where ocean lithosphere plunges back into the mantle, undergoes partial melting and produces *andesitic* magmas (e.g. volcanoes and geothermal centres in the Western Americas and New Zealand). These so-called *hyperthermal areas*, where steam at temperatures well above 100 °C can be brought to the surface for electricity generation, have been the prime targets for past geothermal developments (Figure 25a). Several other countries have *semithermal areas*, where hot groundwater below the boiling temperature can be exploited. But by far the greatest part of the Earth's crust is 'non-thermal', as a geothermal resource. Nevertheless, even in non-thermal areas, useful hot groundwaters and/or exploitable hot dry rocks can exist.

To understand the origin of the different kinds of geothermal resource you need to know the following equation:

$$q = \frac{k\Delta T}{z} \qquad (4)$$

where the *heat flow, q,* is measured in W m^{-2}, the *thermal conductivity, k,* is measured in W m^{-1} K^{-1} and the temperature difference, ΔT, measured over depth z comprises the temperature, or *geothermal gradient*, which has units of K m^{-1} (notice that the SI temperature unit, the Kelvin, is used by convention rather than degrees centigrade). The concept of thermal conductivity requires

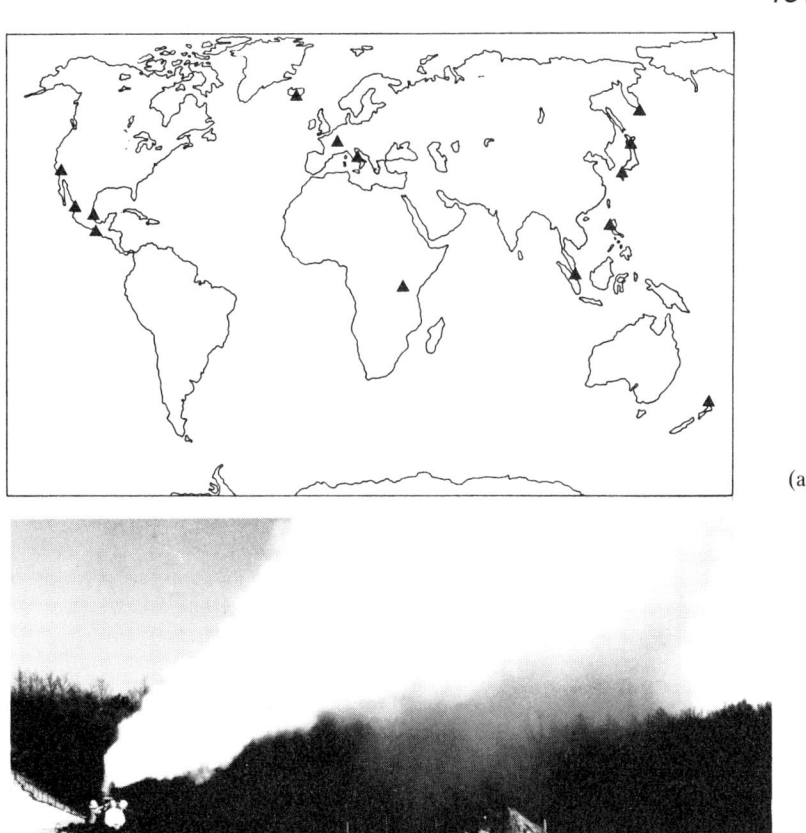

(a)

(b)

(c)

Figure 25(a) The main areas of geothermal energy production operating today; (b) steam at about 5 atmospheres pressure and 150°C being vented from an aquifer at 2 km depth beneath a geothermal site in northern Italy; note condensation over forest area to rear; (c) the Larderello geothermal power station which receives superheated high pressure steam from well-heads (foreground) via steel-pipes covered in lagging and surrounded by an outer aluminium sheath.

further comment. It is a measure of the ease with which a material transfers heat. Thus a metal pan has a high thermal conductivity whereas a pair of insulating gloves has a low value. All rocks are poor conductors of heat in the everyday sense but certain rocks, such as sandstones and granites, are better conductors of heat than shales and many metamorphic rocks, which are the best insulators.

> According to equation 4, if the heat flow has a given (constant) value, will the temperature increase more rapidly with depth if the thermal conductivity is high or low?

If q is constant, a higher value of k will produce a lower value of $\Delta T/z$ and vice versa. Since the question asks for a rapid increase of temperatue (T) with depth (z) a *low* thermal conductivity is required. It follows that if mudstones or shales are present within the top few kilometres, then the rate of increase of temperature with depth will be greater than if high conductivity strata are present (at constant q). Other rock types which can produce an above-average thermal gradient, this time by virtue of their anomalously high concentrations of radioactive heat-producing elements (e.g. uranium, Section 1.2), are granite intrusions. Local heat flow anomalies might occur over such intrusions owing to the additional heat they add to the average, or background, heat flowing through the crust.

You now have the key to understanding the three main categories in geothermal energy exploitation which are considered below. In chronological order of their development these are: (i) hyper-thermal volcanic zones, (ii) non-thermal and semi-thermal zones with shallow, low conductivity strata, (iii) non-thermal and semi-thermal zones characterized by rocks with high heat production. Resources of category (ii) usually include aquifers with warm water within the low conductivity sequence, and this water is exploited by pumping, whereas those of category (iii) are usually dry and must be fractured to allow heat to be extracted by pumping water from the surface down through the rocks and back up to the surface.

2.1   Hyper-thermal resources

Most areas containing recently active volcanoes and/or igneous intrusions will have above-average heat flow and geothermal gradients simply because hot magmas have risen high into the crust and are still cooling. If such an area contains rock sequences which are permeable to groundwater circulation, then the heat will promote *hydrothermal convection* within these sequences, causing some of the groundwater to circulate towards the surface, in turn producing hot springs, boiling mud pools and geysers. Since volcanic activity in a given region usually lasts for a few millions of years and intruded magmas take even longer to cool, hyper-thermal areas effectively provide a *renewable* energy resource because the water supply is also renewable.

You will recall from Part I, Section 7 that oil and gas fields consist of traps, in which hydrocarbons have collected beneath impermeable cover rocks. Analogous requirements apply to geothermal resources if they are to be used for power production: *as well as a heat source, permeable aquifers overlain by impermeable seals are essential.* In fact, the analogy is even closer, for if the water is at or above its boiling temperature it may separate, or *flash*, into steam and liquid water (which will behave rather like oil and gas in petroleum traps) when pressure is reduced by drilling a borehole into the reservoir. Typically, goethermal water may be under a pressure of several tens of atmospheres and at a temperature of 200°C when trapped: steam is produced when the pressure is released. Water extracted during production, and from hot springs and geysers, is usually replenished by water re-entering the aquifer from a *recharge* area.

Figure 26 is a cross-section through the Wairakei geothermal field in New Zealand, which shows the way in which heated water travels upwards, confined at pressure within permeable layers. Faults in the area act as safety valves for the system: the steam and hot water which escape along these outlets are responsible for the surface hot springs and geysers. At Wairakei, steam migrating upwards mixes with cooler waters near the surfce and some condenses: this type of geothermal field is termed 'water dominated' and the steam must be 'dried' to remove any liquid water before it enters the turbines.

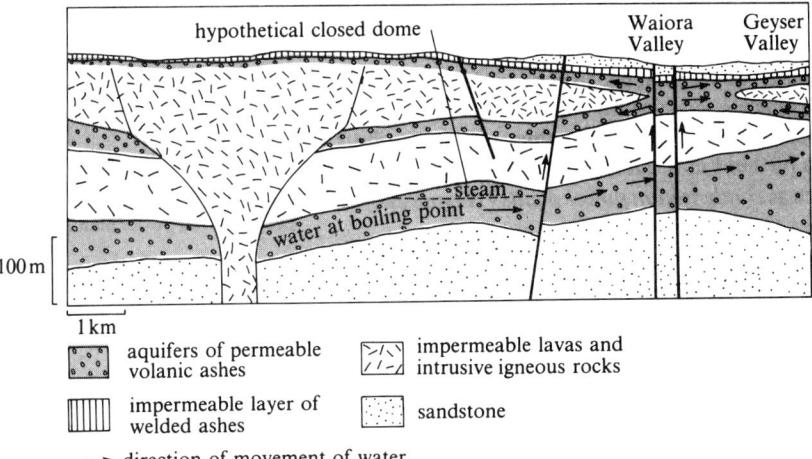

Figure 26 Diagrammatic cross-section of the Wairakei geothermal field in New Zealand showing deep aquifers carrying boiling water, heated by an igneous intrusion, towards the Waiora Valley where faults provide a channel by which the super-heated water (above the atmospheric boiling temperature) reaches the surface as steam. (Note the presence of steam in a dome where upwards circulating water at boiling point is converted into steam due to a lowering of pressure.)

In contrast, the geothermal field at Larderello in northern Italy, one of the world's largest and longest established areas of geothermal power production, is 'steam dominated' or 'dry' (i.e. the steam is free from liquid water) mainly because the base of the impermeable seal is domed and there is no mixing with cooler, shallower water. The Larderello area has been producing electrical power since 1904 from steam circulating within a permeable fractured limestone which is overlain by impermeable shale and clays (Figure 27). The main source of heat is believed to be a magma body and/or a series of ingenous intrusions, several kilometres below the surface. As the intrusions rose through the crust, the area became gently domed and faulted to provide a rather complex *anticlinal trap* for hot water and steam.

Although the principles of generating electricity in hyper-thermal areas are simple enough, there are some formidable technical problems in drilling down to tap steam and hot water at pressures up to 30–35 atm and temperatures up to 400°C. Even at the more typical 4–5 atm at 200°C, high pressure steam emerges from boreholes with velocities of around 1000km h^{-1} (Figure 25b) and sounds like a jet engine at full throttle. This energy has to be transmitted to generator

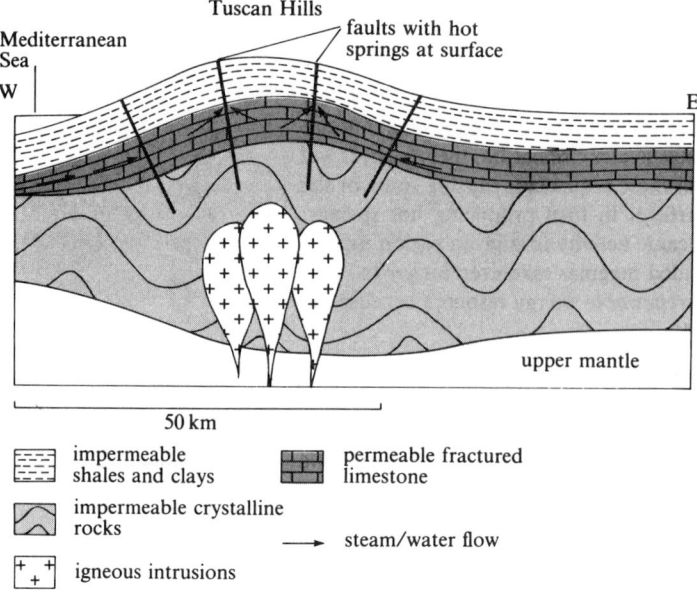

Figure 27 Schematic cross-section through the Larderello geothermal field.

turbines with minimum loss and so the transmission pipes are well lagged to stop the steam condensing to form water droplets and so keep it 'dry'. During the early stages of development, steam from a well is usually passed through a turbine and vented directly to the atmosphere (often via a gas extraction unit which removes pollutants such as carbon dioxide, hydrogen sulphide and methane). Power stations that vent *steam* at just over 1 atm pressure are known as *back-pressure power plants* (Figure 28a) because they rely entirely on the excess pressure of the geothermal steam, above 1 atm, for power production.

Once the output of a geothermal well has stabilized, it becomes possible to increase the efficiency of the turbine by attaching a condensation unit to the low-pressure side. This condenses the post-turbine steam into water, thus increasing the pressure drop across the turbine by a further 1 atm. The power stations of this type require cooling towers and are known as *condensation power plants* (Figure 28b), and the Larderello Station (Figure 25c) is of this type.

A typical turbine produces 15 MW of electrical power and there might be up to four turbines in a power station. In the mid-1980s there were 17 geothermal stations in north-western Italy alone, with an output capacity of 420 MW, or 1.5 per cent of Italian electrical energy demand. 55 MW turbines have been developed in the USA and, by 1985, the Geysers field of N. California was producing 1500 MW of electrical power from only 18 stations, each typically with two turbine units.

A complication in many of these geothermal areas arises from the powerful solvent action of hot aqueous fluids on the rocks through which they pass. As a result, the fluids often become highly saline and can have a severe corrosive effect on machinery. To minimize the long-term storage of these corrosive fluids and to protect the surface environment from pollution, the condensed fluids are usually re-injected into the aquifer at a point remote from the production well.

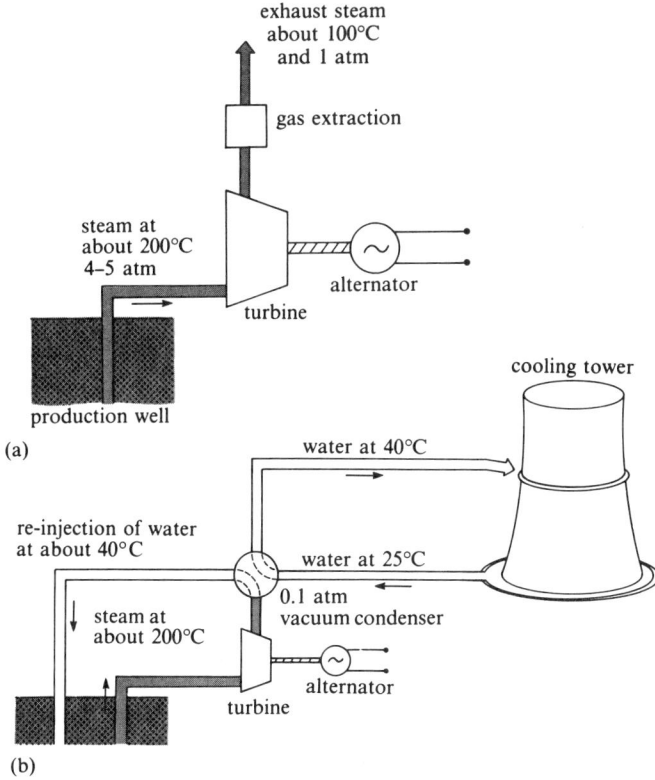

Figure 28 Schematic illustrations showing the operation of (a) a back-pressure geothermal power plant and (b) a condensation geothermal power plant in which steam on the low-pressure side of the generating turbine is condensed to increase the pressure difference across the turbine. (Shaded pipe volumes contain steam.)

This has the additional advantage of replenishing the aquifer with water which may, during migration through the aquifer, be converted back into steam for later production.

This raises one important qualification about the renewable nature of hyper-thermal geothermal resources in the form we have described: the rate of removal of hot water and steam from the system must not exceed the rate of replenishment of heat or water to the aquifer. If the rate of water withdrawal is too great, the pressure falls in much the same way as it does in oil and gas fields nearing exhaustion. Already there are signs that a marked decline in geyser activity in North Island, New Zealand, might be connected with excessive geothermal production.

Despite the complications discussed above, electrical power is being produced at competitive costs from many hyper-thermal areas and many others are the subject of exploration. Table 4a summarizes past and projected future output around the world: it is clear that rapid growth in the exploitation of hyper-thermal resources took place during the 1970s and seems likely to continue for the remainder of this century. Between 1969 and 1982 five countries (Philippines, El Salvador, Kenya, Guadaloupe and Indonesia) joined the geothermal league which hitherto had been dominated by Italy, New Zealand and the USA. Others, such as the Central America states (Nicaragua,

Table 4a Distribution and output of geothermal power from major countries with hyper-thermal resources

	Capacity in 1969/MW	Capacity in 1982/MW	Estimated capacity in 2000/MW
Central American states (excluding El Salvador)	0	0	700
El Salvador	0	95	200
Guadaloupe	0	5	30
Iceland	17	63	500
Indonesia	0	30	>50
Italy	390	460	800
Japan	33	290	>10 000
Kenya	0	30	90
Mexico	4	153	>1 500
New Zealand	170	300	1 400
Philippines	0	425	>500
Taiwan	0	0	200
Turkey	0	1	10
USA	83	c.1 200 *	20 000
USSR	3	6	20
TOTALS	700	3 058	>36 000

* A new 110 MW, two-turbine station is added to the geothermal capacity of the Geysers field almost every year; the total output by 1985 exceeded 1500 MW. There are no other significant changes to this table.

Table 4b Non-electrical applications of geothermal resources (1980) (all data in MW equivalent)

Country	Space heating/ cooling	Agriculture/ aquaculture	Industrial processes
France	14	1	—
Hungary	300	370	—
Iceland	680	40	50
Italy	50	5	20
Japan	10	30	5
New Zealand	50	10	150
USA	75	5	5
USSR	120	5 100	—
Others	10	10	5
Total	1 309	5 570	235

Costa Rica, Guatemala, Honduras and Panama) and Taiwan are destined to do so before long. But future increases in world output are likely to be dominated by the USA and Japan, followed by Mexico, New Zealand and Iceland.

To put the electrical power output from hyper-thermal resources into context, bear in mind (from Figure 4, Part I), that in the years for which data are supplied in Table 4a the following world energy production was/will be required: – 1969 – 2×10^{20} J yr^{-1}; 1982 – 3×10^{20} J yr^{-1}; 2000 – 4.6×10^{20} J yr^{-1}. Geothermal contributions to these totals are 1969 – 700 MW (*ca.* 0.01 per cent); 1982–3000 MW (*ca.* 0.03 per cent); 2000 – 36000 MW (*ca* 0.25 per cent). So hyper-thermal resources are providing a small, but rapidly increasing proportion of world energy production: the growth rate in their exploitation far exceeds those of most conventional energy resources.

The contribution of hyper-thermal geothermal resources to world energy production might seem like a negligible 'drop in the ocean'. But you should remember that Figure 4 in Part I gives *total* energy rather than electrical energy production. Table 4b lists the present non-electrical uses of geothermal resources which, overall (including exploitation of non-hyper-thermal areas discussed below), exceed the total electrical output given in Table 4a. Many of the non-electrical applications do occur in hyper-thermal areas (Iceland, Italy, Japan, New Zealand), and it follows that the ultimate energy resource potential of hyper-thermal regions may be an order of magnitude greater even than the figure for the year 2000 given in Table 4, viz. about 360000 MW, or 1.1×10^{19} J yr^{-1}. Even this is only 1 per cent of the heat escaping from the Earth each year (c. 10^{21} J yr^{-1}, cf. Part I, Figure 1).

To summarize, the overall contribution of geothermal energy is already greater than the power output given for 1982 in Table 4a for two reasons:

1   Steam and hot water in hyper-thermal regions are often used for *district heating* schemes or domestic hot water as well a power production. For example, today, well over half the population of Iceland gets its domestic hot water from geothermal sources, and the original fields in Reykjavik are still producing at undiminished capacity after 50 years. The cost to the Icelandic consumer of piped geothermal heating is only 25–30 per cent of the cost of equivalent oil-fired heating. Indeed, Iceland now derives over 70 per cent of its total energy requirements from geothermal sources.

2   Important contributions to the saving of other forms of energy are already being made by the relatively new technologies, described in Sections 2.2 and 2.3, which are being applied in semi-thermal and non-thermal areas.

2.2   Zones with shallow, low conductivity strata

Many of the countries listed in Table 4b exploit natural flows of hot water for space heating, agricultural propagation etc. Recently, schemes have been devised to extract sub-surface heat in areas where such flows do not occur. Earlier, we noted that in semi-thermal and non-thermal regions there are two possible reasons for above-average geothermal gradients:

(i) the top few kilometres of crust may have better insulating properties than average — sedimentary basins are ideal, and (ii) the upper crust may contain significant concentrations of heat-producing radioactive elements. This Section examines the first of these characteristics with particular reference to developments in the Paris area during the late 1970s and early 1980s. In this type of non-volcanic geothermal system, heat is extracted faster than it is replaced; *it is therefore a non-renewable energy resource–heat energy is being 'mined'*.

Beneath the city of Paris and the surrounding area lies a large basin 500 km wide and 1–2 km deep, of Mesozoic (Traissic to Cretaceous) sedimentary rocks, some of which are important aquifers. Beds of coarse-grained porous limestones and sandstones alternate with impermeable clays, shales and marls (calcareous clays), as shown schematically in Figure 29.

> Apart form its impermeable nature, in what other way is the clay–shale–marl group important geothermally?

These are the rocks with low thermal conductivity across beds of which high geothermal gradients become established. So the coarser grained bands of limestone and sandstone act as aquifers and carry meteoric water to depth where it is heated by a combination of normal heat flow and enhanced temperatures due to the insulating effects of the overlying low conductivity strata; the latter also seal water into the aquifers. Several horticultural schemes (e.g. at Melleray near Orleans) and district heating schemes (e.g. at Creil, north of Paris; Figure 29) have come into operation since 1978. They use water at between 55 and

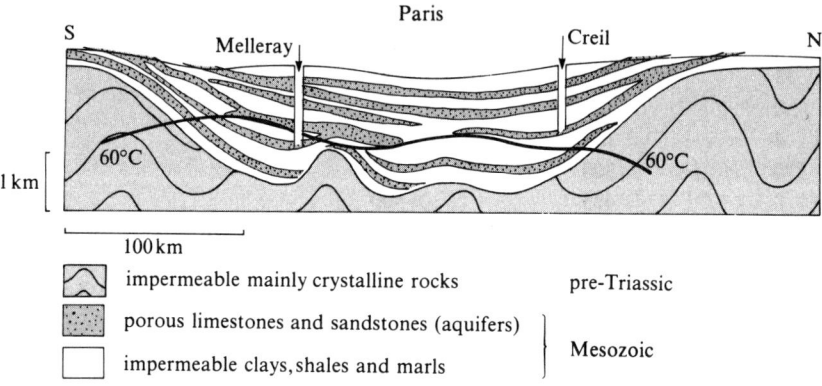

Figure 29 Cross-section through the Paris basin showing the approximate positions of buried aquifers which alternate with less permeable, insulating layers of clay-rich sedimentary rocks. The 60°C isotherm is marked, and the approximate positions of the two geothermal installations discussed in the text are also indicated.

70°C which either rises under the natural head of *piezometric pressure*, or is pumped, from 1–2 km depth. The hot water is highly saline and too corrosive to be allowed directly into heating systems, so it is passed through corrosion-resistant heat exchangers (Figure 30a). Here the geothermal heat is transferred to freshwater circulation systems connected to the distribution circuit, which incorporates the means of using the heat, known as *heat load*. The secondary system in the vast greenhouse complex at Melleray consists of both aerial and undersoil pipes. In contrast, a typical domestic heat load, as at Creil, might comprise a combination of underfloor and radiator pipes installed in 3000 –4000 apartments in large blocks of flats (Figure 30b). After passing through the heat exchanger, the cooled saline water in the primary geothermal circuit, now at 20–25°C, is re-injected into the aquifer, normally at a point about 1 km away from the production well. This maintains the pressure in the aquifer and minimizes pollution.

At the design stage, a twin production and reinjection borehole system would be planned on the basis of supplying between 5 and 10 MW of heat energy over a lifetime of 30–50 years. After this time, the heat in the aquifer at the base of the

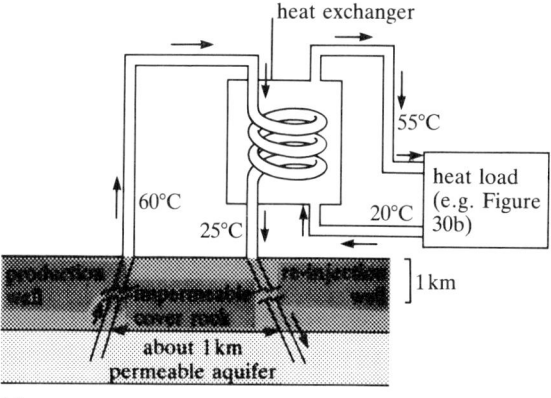

(a)

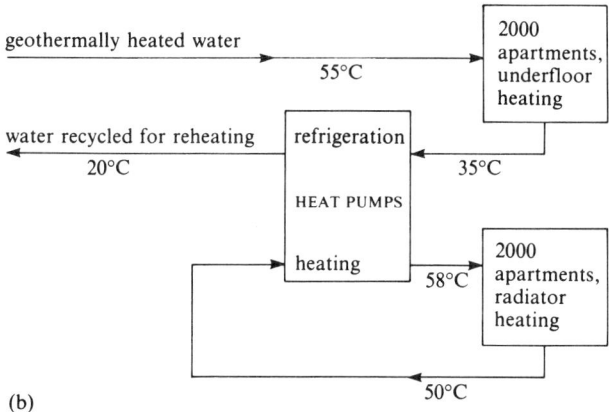

(b)

Figure 30 Simplified scheme for domestic or industrial space heating as used in the Paris area. In (a), heat is exchanged between the geothermal water circuit and the secondary, freshwater circuit. In (b) the water in the secondary circuit is used first to heat 2 000 apartments by underfloor heating; the residual heat energy is then transferred to a smaller volume of water using heat pumps (see text) and is used for radiator heating in a further 2 000 apartments.

production well would be depleted and would take several hundred years to be restored. Because the capital costs of a single installation are high (c. £2.5–3 million), it is important to plan for a consistently high heat load during its lifetime. Overall, the economics of these schemes look very attractive: the cost has proved to be slightly less than the cost of comparable oil-fired systems. More important still, each twin borehole installation, of which five were in production in the Paris area during the early 1980s with more planned, provides an annual saving of 3 000–4 000 tonnes (20 000–30 000 barrels, or nearly a million gallons) of oil that would otherwise be used to provide the necessary heating. Indeed, many of the geothermal installations do have oil-fired back-up boilers for use in winter.

One reason why the French took the lead in developing the technique of exploiting geothermal heat from sedimentary basins was because the geology of the Paris basin was already well known from the many oil exploration boreholes that had been drilled (Part I, Figure 32). Success with geothermal heating in this sedimentary basin has led to interest elsewhere, particularly in European countries where exploration for sites with suitable geological characteristics is well advanced. In the UK attention has focused on two deep basins, which also contain oil (cf. Part I, Figure 47), centred on the Dorset–Hampshire and east Yorkshire–Lincolnshire areas. Exploratory geothermal wells in the Southampton area and at Cleethorpes, drilled during the early 1980s, intersected Permo-Triassic sandstone aquifers at about 2 km depth but generated, respectively, lower sustainable flow rates and lower temperature water than had been predicted. These initially disappointing results, combined with the fact that most UK households are heated independently and that the competing fuels – mainly natural gas – have not increased in price as anticipated, weaken the case for widespread development of geothermal aquifers. Unlike France, where the main substitute fuel is oil; in the UK such schemes do not look economic at current (early 1986) fuel prices though they may become so in the future if fuel prices rise in real terms. Nevertheless, most assessments of energy supply in the UK covering the next 20–30 years indicate the geothermal aquifers will have a minor role to play in the provision of low grade heating where purpose-built heating systems can be incorporated into new building designs. A contribution equivalent to about 200 MW (ca. 6×10^{15} J yr^{-1}) in the UK by the year 2025 with an upper limit of 500 MW, has been estimated for this energy resource by the UK Department of Energy.

Finally, it is interesting to note the potential of heat pumps to enhance the efficiency of geothermal systems. A *heat pump* performs the apparently magical task of taking thermal energy from a dilute source (e.g. a large volume of water at a low temperature) and concentrating it into a useful form (e.g. a small volume of water at a high temperature). Heat pumps are simply the more familiar air conditioner operating in reverse. In hot climates, air conditioners take heat from inside a building and dump it outside at a higher temperature whilst refrigerating the interior. The installation at Creil (Figure 30b) uses geothermally heated water at 55 °C to supply underfloor heating to 2000 apartments, by which stage the water has cooled to 35 °C. Intermittent operation of heat pumps concentrates the energy from this water and transfers it to water in a separate circuit supplying radiator heating to a further 2000 apartments at 58 °C. The overall efficiency of the system is therefore critically dependent on the action of heat pumps to concentrate the remaining thermal energy after the first set of apartments has been heated.

In the longer term, heat pumps may allow the commercial development of shallower, cooler geothermal aquifers. The lower cost of drilling to shallower levels combined with the widespread availability of shallow aquifers and the ability of heat pumps to produce small volumes of water at useful temperatures, may bring the cost of geothermal heating within range for small groups of dwellings. Clearly, owing to the nature of housing developments, this could enhance the prospects for geothermal district heating schemes in the UK.

2.3 Hot dry rocks

Although the temperature gradient within a sedimentary basin, such as in the uppermost 1–2 km of the crust beneath Paris, may be above average, the gradient will be lower beneath the sedimentary layers where crystalline rocks (of normal thermal conductivity) are present. But if these rocks have high heat production then there is a good chance that high geothermal gradients might persist to much greater depths. Beneath the granite moorlands of south-west England, for example, it is estimated that temperatures within granite intrusions rich in potassium, uranium and thorium may reach 180–190°C at 5 km depth — higher temperatures than are likely to occur at the same depth beneath Paris. Of course, these high temperatures are the result of radioactive decay rather than of magmatic heat in these 270 Ma old granites. Such temperatures are interesting because of their potential for raising steam to drive turbines and produce electricity. South-west England is typical of several regions where above-average surface heat flow testifies to immense geothermal energy resources in *hot dry rocks (HDR)* that lie within the range of deep drilling. The problem is that, whilst these rocks are *hot* because of their radioactive heat production, they are also *dry* because they are impermeable, and so are unable to carry a working fluid (i.e. water) by which their heat content could be exploited.

There have been two significant experimental attempts to overcome the formidable technological obstacle of creating a heat exchange surface at depth in hot dry rocks through which the working fluid can be passed. The first of these was initiated in the mid-1970s by the Los Alamos Scientific Laboratory (LASL) in New Mexico, USA; it consists of pumping water down a borehole, drilled to depths where there are steam-raising temperatures, at progressively increasing pressure until the rock at the base of the borehole splits to form a narrow vertical crack — a technique used in the oil industry and known and *hydro-fracturing*. A second borehole is drilled to intersect this crack several hundred metres above its base. A closed circuit for water circulation is established in this way by pumping water down the first hole and extracting steam from the second, the large area of fracture providing a good surface for heat exchange. The main problem with this technique is that hydro-fracturing produces near the hole only a *single* crack through which all the circulating water must pass. Much power is needed to force the circulating water through this single crack; nevertheless, the LASL team has formed suitable fractures at 3 km depth in granite and has conducted extensive tests, producing steam at 200°C for power generation. Note however, that New Mexico is a semi-thermal region with some active volcanoes and therefore has a higher geothermal gradient (*ca.* 60 K km^{-1}) than that in the granites of south-west England (*ca.*37 K km^{-1}).

More recently, during the early 1980s, workers at the Camborne School of Mines (in Cornwall) have developed a technique for obtaining access to the deep natural fissures in the granite intrusions of south-west England. Twin boreholes are drilled and are deviated at depth to intersect natural rock joints and fissures (Figure 31); a vertical spacing of several hundred metres is left between the two holes. Small explosive charges are detonated in the lower hole (stars in Figure 31b) to initiate a multiple fracture system around the hole. This provides many passageways by which fluids may gain access to the natural fissures. Finally, water is pumped down the lower hole at increasing pressure until the fissures are opened hydraulically to provide a water circulation circuit. The resistance to water flow is much less in this multiplicity of cracks, which allow freer movement of water, than in the single crack achieved with hydro-fracturing alone. This 'explosive stimulation' technique has been shown to provide the required low resistance to water flow in experiments carried out in the granite initially at a depth of 300 metres (in 1981) and subsequently at 2000 metres (in 1983–5) though initial tests were hampered by downwards loss of water, presumably into altered granite beneath the zone of fracturing. If a commercial HDR system is developed in Cornwall this technique will need to be carried out in boreholes 5–6 km deep, from which steam at 150–200°C could be produced.

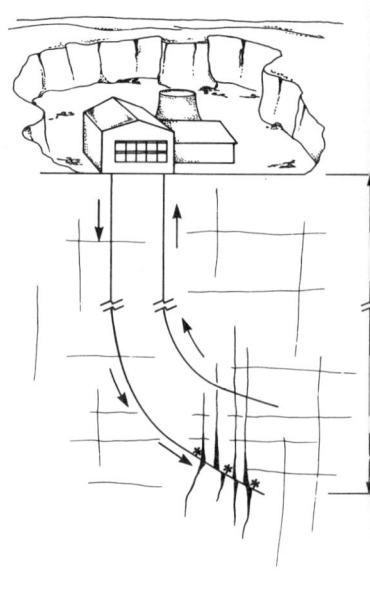

(a)

Figure 31(a)   Drilling rig on-site at the Rosemanowes Quarry location where boreholes to 2 km depth were sunk for HDR rock mechanics tests during 1983–4, (b) Aspects of the underground circuit being developed in south-west England for the extraction of geothermal heat from hot dry rocks.

Although the HDR technologies remain to be proved in commercial terms, all the pilot experiments on steam and power generation at Los Alamos and on rock mechanics in Cornwall have so far been successful. Calculations indicate that geothermal electricity costs in the UK would lie at an attractive 5 pence per kWh, including writing off the costs of drilling the geothermal wells over their anticipated 30 year lifetime. Although, like the sedimentary basin geothermal systems, HDR developments involve the mining of heat, they have the significant advantage of producing electical power which can be transmitted to distant markets.

South-west England is a non-thermal area with heat flow that is only slightly above average, yet it has been estimated that the HDR geothermal resource base in this area alone might be of the order of 8000 million tonnes of coal equivalent — more energy than a recent British Coal estimate of coal reserves, 7000 million tonnes (Part I, Section 4.7). Although it is unlikely that HDR sytems will be developed widely enough to extract much of this energy in the foreseeable future, such calculations emphasize the potential importance of such systems. In the longer term, as technology improves, a much wider variety of less favoured geological sites could be exploited by HDR methods. The conclusion is that, given the necessary investment and technology, geothermal energy resources could increase their contribution from a fraction of one per cent, at least to several per cent of world energy supply by the middle of the twenty-first century.

2.4 Summary of Chapter 2

1 Traditionally, 'renewable' geothermal energy has been exploited by drilling for steam at high pressure and high temperature (up to tens of atmospheres pressure and up to 400°C) from aquifers in hyper-thermal areas along active plate margins (see Table 4).

2 Electricity production from hyper-thermal geothermal resources was about 3000 MW (1×10^{17} J yr^{-1}) in 1982 and is planned to exceed 36000 MW (1.1×10^{18} J yr^{-1}) by the year 2000. Additional savings in other energy resources result from the piped distribution of geothermal steam through district heating networks to both domestic and industrial consumers.

3 New developments to exploit 'non-renewable' geothermal resources in semi-thermal and non-thermal areas include: (i) pumping hot water, again for domestic and industrial space heating, from aquifers in sedimentary basins where temperature gradients are enhanced owing to the presence of low conductivity rocks in the sedimentary sequence; (ii) creating deep fractures and water circulation circuits by explosive stimulation and hydro-fracturing of hot dry rocks, to produce steam for power generation from granites with enhanced heat production.

4 Given appropriate investment and technological developments, together with further increases in production from 'traditional' geothermal areas, these new resources have the potential to bring the geothermal contribution to world energy supply above 10^{19} J yr^{-1} by the middle of the twenty-first century.

3 Surface energy resources

Study Comment. In this Chapter we consider renewable energy resources available at the Earth's surface, viz. those that depend on solar radiation alone (solar and biomass energy), or in combination with the Earth's rotation or its gravity (wind, wave and hydroelectric power) and on the gravitational effect of the Sun and Moon (tidal power). As you read this Chapter you may find it useful to refer to Figure 1 in Part I, which shows the main sources and directions of energy flow on the Earth surface.

It should be clear by now that economic grades of secondary energy resources (i.e. fossil and nuclear fuels) are sufficiently abundant to fulfil our energy requirements for the next few decades. However, several forms of *primary energy* (i.e. renewable energy income) are likely to play an increasingly important role in reducing the demand for these non-renewable fuels. Many renewable forms of surface energy, often termed 'alternative energy resources', which include hyper-thermal geothermal resources, have been used in various forms for centuries. For example, the *kinetic energy* of moving air and water has been harnessed throughout much of recorded history. At their peak, there were probably 20000 water mills operating in Britain for milling, grinding, weaving, etc., whilst windmills were a feature of the agricultural landscape for many generations. But unlike fossil fuels, these energy sources are not transportable and were available for use only near the site of conversion, which is not ideal for urban, domestic and industrial uses, many of which require electrical power. Also, they are mainly fluctuating energy sources, due either to diurnal fluctuations (solar and tides), or random fluctuations (wind, waves, etc.). In more leisurely times, it was largely acceptable to wait; today, this does not apply and the technology of renewable alternative energy resources is based on either (i) finding a long-term storage method with minimum energy loss, or (ii) accepting the random fluctuations and using the energy source as a back-up to reduce the consumption of non-renewable fuels. Among the most predictable sources, hydropower (Section 3.3) already provides about 4 per cent of global energy supply, and tidal power (Section 3.4) is seen as a natural extension of hydropower technology. The less predictable sources of wind and wave energy (Sections 3.2 and 3.5) are also receiving serious consideration in Britain. Overseas there is considerable interest in developing various techniques for converting solar energy into electricity and biomass into transportable fuels (Section 3.1).

3.1 Solar energy

At present, *solar energy* contributes a negligible amount to world energy supply, yet only a small fraction of the incident solar radiation would need to be trapped in order to satisfy total demand. The main problems are (i) the intermittent, dispersed and seasonal nature of solar radiation, which makes its recovery difficult and expensive, particularly in the UK; (ii) the large amounts of land required by and the low efficiency of solar power converters; and (iii) the long-term storage required if solar energy is not converted into electrical power.

Table 5 illustrates the first point — the wide seasonal variation in solar input (i.e. the average incident radiation) in the UK and Ireland compared with five other countries that have active solar energy programmes. Table 5 shows that although the annual mean solar input in the UK is only a factor of two or three less than that of countries in low latitudes, the solar input per head of population, and the ratio of solar input to total energy use are much lower than for those countries, and even lag well behind the data for Ireland. Variations in the last two columns of Table 5 are due, first, to population density which is much greater in the UK, Japan, India and Israel than it is in Ireland, the USA and particularly Australia. Secondly, primary energy use per head of population is much greater in some countries than in others. Thus, for example, the USA has a large solar input per head and a large energy use per head (ratio 950) whereas India has a small solar input per head but an even smaller energy use per head (ratio 8000). So the UK has the least advantageous combination of a small solar input per head and a large energy use per head and, therefore, has the lowest ratio of solar input to total demand among the countries listed. However, some of the methods used to exploit solar energy elsewhere are amenable to development in an energy saving context throughout high latitude countries, such as the UK. Figure 32 summarizes the major solar energy technologies that are being exploited on a world scale. The principles of each technique and its wider applications are discussed in Sections 3.1.1–3.1.3; finally, Section 3.1.4 gives a brief introduction to the potential for solar energy exploitation in the UK.

Table 5   Solar input statistics for various countries

Country	Energy input/ 10^6 J m^{-2} day^{-1}			Energy input per head/ 10^{12} J yr^{-1}	Ratio of annual solar input to annual energy use
	midsummer	midwinter	annual mean		
United Kingdom	18	1.7	8.9	12	100
Ireland	18	2.1	10	85	790
Japan	17	7.0	13	16	160
USA	26	11	19	310	950
Australia	23	13	20	4 300	28 000
India	26	14	20	44	8 000
Israel	31	11	22	60	680

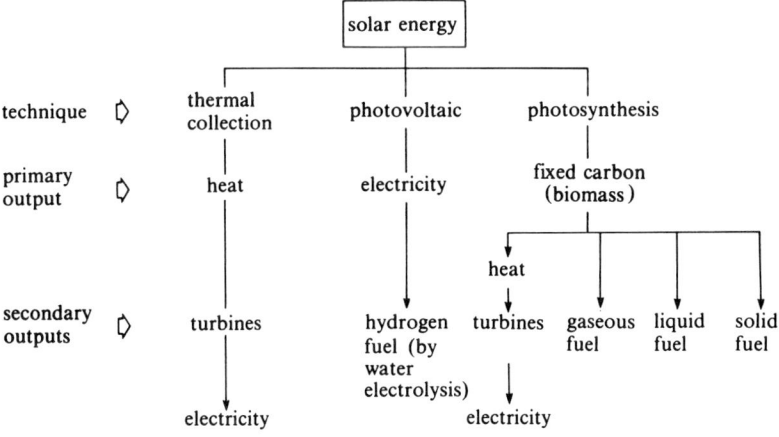

Figure 32   The three principal routes in the exploitation of solar energy.

3.1.1 Thermal collection of solar energy

Thermal collectors are of two types: (i) flat-plate collectors, which are used to generate heat only (cf. Figure 32) for domestic and industrial space and water heating; and (ii) focusing collectors, which are used to operate solar furnaces at high temperatures for electrical power generation.

Flat-plate collectors

You only have to visit an unventilated greenhouse on a sunny afternoon to apreciate that most buildings with glazed windows absorb solar energy. The interception of solar energy may be maximized by improving the absorptive properties of the collector and by orienting it at the most effective angle to the Sun. Collectors therefore usually incorporate a glazed surface underlain by copper or aluminium tubes filled with air or water, which are connected to a system for water or space heating.

Design studies and practical applications to date show that a $1\,m^2$ solar collector panel has an average output of 1.3×10^9 J yr^{-1} and that, for a typical house, 4–5 m^2 of panels are required for water heating. A much larger area, say 40 m^2, equivalent to all the south-facing roof area of a typical house in the northern hemisphere, would be required to produce enough energy for space heating. Nevertheless, since domestic space heating accounts for nearly 20 per cent of total energy demand in temperate latitudes, the potential savings are enormous. But there are two snags: first, the large capital cost of solar collectors (about £300 m^{-2} in the early 1980s), though these would be less if the collectors were incorporated during house construction; secondly, there is the problem of energy storage from the period of peak summer input to the peak winter demand. Some ingenious solutions have been devised, such as collecting the heat in underground water tanks or rock stores. however, it is difficult to assess the long-term contribution to global energy supply that flat-plate solar collectors might make; most of the savings in the foreseeable future are likely to be small and due to the enterprise of individuals.

Focusing collectors

These are designed to concentrate the available solar energy to produce a secondary output at high temperature (Figure 32) for power generation. Some designs use an array of sun-tracking mirrors (heliostats), whereas others have a system of lenses, focused on a solar furnace where temperatures of several thousand degrees centigrade are achieved. Superheated steam produced in the furnace is forced through a generating turbine and is then condensed to water ready for recycling (Figure 33).

Figure 33(a) Schematic illustration showing the operation of a heliostat–solar furnace power plant. (b) The 10 MW experimental solar-power plant '*Solar-One*' in California's Mojave Desert where enough electricity is generated to run the plant and meet the needs of 6000 people. This view, looking west, shows a portion of the 72-acre heliostat field with 1240 heliostats in the north of the field to the right and 578 in the south. The heliostats are stowed in the face-down position in this picutre. (c) The central tower of Solar-One has a black surface which glows bright white under reflected, concentrated sunlight. The receiver is 7 meters in diameter and contains 24 vertically-oriented nickel alloy steel tubes through which water passes and 'flashes' to steam at 500°C for electricity generation in turbines. This plant also contains a turbine by-pass which allows heat energy to be passed to a special oil for storage; the thermal storage system can reproduce steam for power generation for four hours during cloudy or night-time conditions.

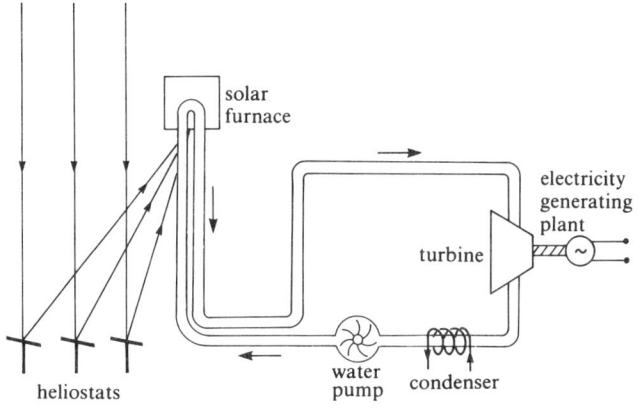

solar
furnace

electricity
generating
plant

turbine

water
pump

condenser

heliostats

(a)

(b)

(c)

Numerous experimental solar power stations with different design features have been developed in western Europe and the USA. The largest European solar furnaces in operation in the early 1980s were at Odeillo in the French Pyrenees and at Adrano, Sicily. They generate only small amounts of electricity (1MW at Odeillo) compared with fossil fuel and nuclear stations. The experimental solar furnace plants in the USA, notably at the Georgia Institute of Technology, at the Sandia Laboratories in New Mexico and in the Mojave Desert of California (Figure 33) have electrical outputs ranging up to 10 MW. But to make a significant contribution of, say, 1 000 MW (3.15×10^{16} J yr^{-1}), the incident solar energy over an area of about 18 km^2 would need to be focused onto a solar furnace. This estimate was based on the optimum conditions near El Paso in the sunny south-western desert of the USA. Even in this location a 40 per cent contribution from fossil fuel boilers is needed to produce 1 000 MW continuously. Taking a country like the USA as a whole, it is estimated that about 6.3×10^{19} J yr^{-1} of electrical power will be required by the year 2010. In the unlikely event that this were to be provided by combined solar-fossil fuel schemes of the type just described we would require:

$$6.3 \times 10^{19}/3.15 \times 10^{16} = 2000 \text{ power stations}$$

occupying an area of $2000 \times 18 = 3.6 \times 10^4$ km^3, which is 0.4 per cent of the US land area. It follows that there is no shortage of land in a country like the USA for developing solar power stations. However, widespread development of solar furnace power stations is unlikely for two reasons. First, the operation would have an enormous demand for aluminium and steel (used in reflectors and their mountings) and second, the capital costs are estimated at 4–5 times greater than those of the equivalent nuclear power station. Although focusing collectors may become increasingly used in low-latitude countries they are unlikely to make a major impact on global energy supply.

3.1.2 Photovoltaic conversion of solar energy

The idea of using a light-sensitive *solar cell* is that it should convert *sunlight*, rather than heat, directly into electricity. The first low-efficiency photovoltaic cells using cuprous oxide on copper were developed for photographic exposure meters before 1940. During the 1950s the Bell Telephone Company introduced a silicon-based cell with a conversion efficiency of 11 per cent which, despite trials with other materials, is now used extensively, even for auxiliary power and communications on most rockets and space vehicles. The basic principle is that a cell consists of flat discs of two types of silicon (with different impurity elements) mounted one on top of the other, and when sunlight falls on the top disc a current flows between the two.

The maximum theoretical output of solar cells in bright sunlight, about 220 W m^{-2}, is rarely achieved at the Earth's surface because some of the critical short-wavelength radiation does not reach the surface. In ideal conditions an *average* output (i.e. on a 24 hour basis and including seasonal variations) of around 30 W m^{-2}, or 30 MW km^{-2}, would be more likely. To generate 1 000 MW using solar cells would require, therefore, 33 km^2 of land area, a similar area to the solar furnace (which requires 18 km^2 to produce 60 per cent of 1 000 MW).

> Can you think of any other problems that might inhibit the widespread commercial development of solar cells?

Apart from the land requirements the main problems are those of energy storage and the cost of solar cell fabrication. Electrolytic dissociation of water to produce hydrogen fuel (Figure 32) may, in the long term, offer an appropriate storage method (Section 5). But at £150 for every square metre of cell area, the manufacturing costs are still ten to twenty times greater than required to be competitive with conventionally produced electricity. Research into cell technology coupled with mass production at cheaper rates is being pursued, however, and there is even the suggestion that large-scale solar power units might be assembled in orbit around the Earth where they could take advantage of uninterrupted, unattenuated sunlight and beam down the power they produce as microwaves.

3.1.3   Energy from the biomass via photosynthesis

Through the photosynthetic fixation of carbon in plants, the Sun provides the energy for all processes in the biosphere. The chemical energy content of plants can be recovered as 'food' energy in the metabolism of animals, or as 'heat' energy in the combustion of biomass materials, e.g. wood, straw, plant matter and other waste organic materials. The process of growing and harvesting energy crops that have trapped solar energy by photosynthesis, for exploitation using one of several *biomass energy conversion* processes, provides a third important method of harvesting solar energy (Figure 32).

> What advantages do you think biomass energy has over the other, more direct methods of solar conversion considered above?

One important advantage is that photosynthesis does not require bright sunlight but continues at a slow and steady rate even if the intensity of sunlight is very low. This means that energy crops could be grown almost anywhere on the Earth's surface though, of course, they would be in severe competition with food crops in densely populated countries at high latitudes. It follows that the potential for producing biomass is greatest in humid, densely vegetated tropical countries, many of which are badly in need of a lucrative export trade.

This brings us to the second, and more obvious advantage of biomass energy — it can be converted into easily transported fuels, as indicated in Figure 32 and summarized below. But why *convert* it into fuels; what is wrong with using biomass materials themselves as fuels?

First, most biomass is of low density compared with coal and oil, and, because much of it is carbohydrate rather than hydrocarbon, its energy density or calorific value, is very much lower than coal or oil (crude oil 36×10^9 J m^{-3}, coal 55×10^9 J m^{-3} compared with wood at 9×10^9 J m^{-3}). Thus large volumes of biomass will be required both as fuel and at the transportation and storage stages. Second, even assuming that biomass is dried before combustion – a process that itself uses energy – there will still be a large residual moisture content which will make it burn less efficiently than coal or oil. Third, because biomass breaks down readily during storage (i.e. it rots) large volumes cannot be stored indefinitely – this is the main reason why biomass is converted into other fuels.

Nevertheless, and despite the problems just discussed, biomass is incinerated for its energy content in many parts of the world, but its potential both for combustion and conversion into fuels is often underestimated, especially by some local authorities in the UK who find it more economic and expedient to bury organic refuse. However, any serious attempt to develop the potential of biomass energy would involve one of the following conversion processes. (i) Controlled burning produces *biogas*, a mixture of methane and carbon dioxide, with a calorific value of 4–5 k Wh m^{-3} (cf. natural gas, about 12 k Wh m^{-3}). (ii)

Pyrolysis, in which organic molecules are strongly heated in the absence of oxygen, yields solid, liquid and gaseous hydrocarbons (Figure 34). Note that the solid fuel produced in pyrolysis, known as *char*, is a carbon-rich substance which has a high calorific value and can be formed into briquettes. (iii) Digestion by bacteria in *anaerobic conditions*, again produces biogas. (iv) Anaerobic *glycolysis* (fermentation) or chemical reduction, in which the organic polymers are broken down, produces liquid alcohols and even hydrocarbon molecules like those in crude oil.

The main advantage of producing such concentrated fuels is that, like conventional hydrocarbon resources, they can be stored indefinitely. Apart from power generation, the main applications of gaseous and solid fuels would be in domestic and industrial space heating and thermally based manufacturing processes. The main potential use of liquid fuels would be in transport; for example, since the late 1970s a large part of the Brazilian sugar crop has been used to manufacture *ethanol*, which is added in a 1:3 ratio to petrol for some cars and trucks. These developments have required relatively small modifications to engine designs. The main practical disadvantages are that ethanol has a lower calorific value than petroleum (7.4 kWh kg^{-1} as against 12.2 kWh kg^{-1}) and, in the early 1980s, is more expensive to produce. Moreover, in Brazil there have been worries about the progressive depletion of food crops in favour of the manufacture of alcohol fuels. But in 1985, following a slump in the world sugar price, the principal sugar exporters in Brazil could not cover their growing costs and were keen to commence a major programme of converting surplus sugar into ethanol in order (a) to stabilise world sugar prices and (b) to reduce ethanol fuel costs below the uncompetitive level of \$52 a barrel (twice the price of crude oil in 1985). Another major attraction of alcohol fuels is that they are relatively clean burning compared with conventional fuels and, for this reason, much

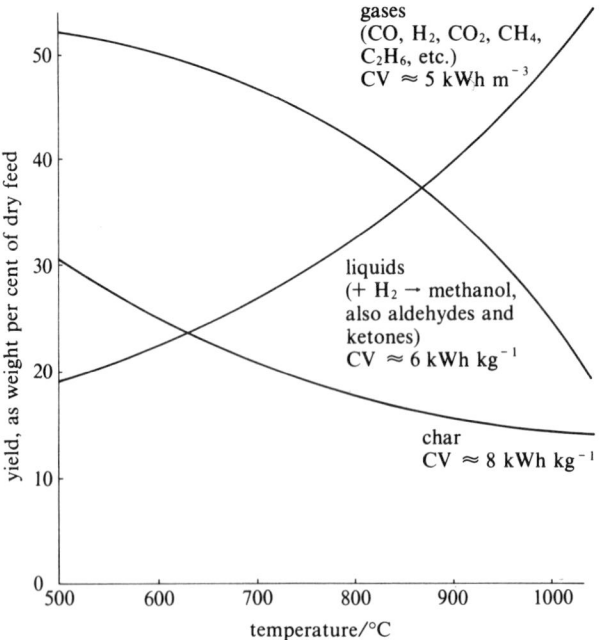

Figure 34 A typical graph of the effect of pyrolysis temperature on the yield of products from biomass conversion. Also shown are the approximate calorific values (CV) of the different products.

research has been carried out into their use in California, traditionally one of the worst-hit smog areas of the world. Work at the University of Santa Clara has provided a technological base for widespread production of methanol-powered vehicles. In addition to the anti-pollution case, it is argued that, as crude oil reserves dwindle, alcohols derived from biomass or coal may become the natural automotive fuels. Thus, by the early 1990s, it is anticipated that many thousands of methanol cars will be operating in California in addition to the *ca.* 1 million pure ethanol cars running already in Brazil. Further developments will depend largely on political factors such as government assessments of the relative importance of reliance on crude oil compared with alternative strategies. Moreover, the alcohol fuel programme will be strongly affected by rapid changes in the price of crude oil such as occurred in 1986.

On a world scale, however, there should be no shortage of potentially available biomass energy — the equivalent of 10^{22} J yr^{-1} is created annually by photosynthesis, whereas energy demand is only about 3×10^{20} J yr^{-1} (Part I, Section 1). Also, major production and widespread transportation of fuels derived from biomass could become economically viable within decades. However, the incentive for these developments would arise only from a major change in energy policy and/or from a worsening of the world supply and demand situation for hydrocarbon fuels and nuclear energy.

3.1.4 Solar energy: a UK perspective

The strong seasonal variation of incident solar radiation in the UK (Table 5) emphasizes the need for a long-term storage facility if solar energy is ever to make a major impact. This is for the fairly obvious reason that the period of peak energy demand is associated with minimum energy input from solar sources. Nevertheless, flat-plate solar collectors which provide low-grade heat for space and water heating are gaining in popularity, though the capital cost of £1500 (1985 prices) for the installation of a 4–5 m^2 area of solar collectors, pumps etc., together with the domestic back-up system for water heating, has inhibited their widespread development. Given that the average UK household spends about £100 a year on water heating and that a separate heating system is needed in winter, solar installations usually take at least 15 years to reach the break-even point. In addition to domestic uses, there is a substantial market for solar heating of swimming pools in the UK and, in industry, commerce and agriculture there are certain specialized applications in which large volumes of heated water are used where solar heating is potentially cost-effective. A recent estimate (Table 6) suggests that the potential exists for some 0.16×10^{18} J yr^{-1}, or 6 million tonnes coal equivalent, to be saved if most future house designs were to incorporate solar water heaters. This is equivalent to 2 per cent of annual UK fossil fuel consumption in the early 1980s. However, in view of their marginal cost-effectiveness solar water heaters are likely only to have limited market penetration in the foreseeable future (ca. 3×10^{15} J yr^{-1} by the year 2025 according to an ETSU energy supply scenario). The prospects for solar water heating in the UK would be significantly enhanced only if there were to be a long-term rise in costs or restriction in availability of conventional fuels, or a major breakthrough in solar technology, in terms of costs or performance.

But what about domestic space heating, which consumes annually 1.6×10^{18} J, equivalent to nearly 20 per cent of annual UK fossil fuel consumption? Theoretically, at least, much of this energy could be supplied from the Sun. For example, the *average annual* solar imput in the UK is about 3.5×10^9 J m^{-2}. The average well-insulated house requires about 6×10^{10} J yr^{-1} for heating so, given an efficiency of about 50 per cent, some 40 m^2 of solar panelling would be required to supply space heating for a typical house. Thus, there is both an adequate supply of solar energy on an annual basis, and there could be an adequate supply of south-facing roof area on most houses, given suitable planning, to accommodate the necessary panels. However, there are two major obstacles which make solar panel space heating even less attractive than water heating. First there is the problem of inbalance between supply and demand, mentioned earlier, which means that heat must be stored for periods of 6–9 months until required. Many experimental schemes exist designed to solve this problem; for example large insulated subterranean water tanks or crushed rock stores (which allow free passage of air during heat delivery and retrieval operations) have received much attention. Nevertheless, the overall efficiency of the installation is reduced by the need for long-term storage and this reduces its cost-effectiveness. The second problem is that space-heating systems are inherently less efficient than water-heating systems because heat is transferred to a large volume of air rather than a small volume of water – and there is the added problem of heat loss on a larger scale. In their 1985 review, ETSU concluded that 'there is no prospect that active solar space heating will become cost-effective in the United Kingdom'.

The schemes considered above are all classified as *'active'* solar designs in that they involve a collector (solar panel) and a heat transfer medium (water). Recently much attention has focused on so-called *passive* solar designs which use the fabric or form of a building to capture sunlight and to reduce the need for heating and lighting from conventional sources. Particular design measures (e.g. Figure 35a) that have been evaluated are:–
(i)   south orientation of the building, with most of the window area on the southern side, preferably double-glazed to optimise the heating effect,
(ii)   roof-collectors consisting of glass panels that allow heat to collect in the loft space for transfer via. ducting fans to the living area,
(iii)   additional conservatory areas which function in a similar way to roof collectors, and
(iv)   special wall designs (e.g. the Trombe wall, named after its designer) which have adjustable louvres and insulating shutters operated by timing devices and which optimise the circulation of hot and cold air throughout the house.

An important factor is that passive solar designs should be seen a part of a total package of energy efficiency measures. In this context, the Department of Energy has shown that these schemes are already cost-effective. For example, for an investment of *ca* £100, some 1000–2000 kWh can be saved annually from heating bills, a situation that reaches breakeven point in only a few years. Recent estimates (Table 6) suggest an ultimate technical potential saving in UK primary energy resources from passive solar designs of 0.43×10^{18} J yr^{-1} with some 60×10^{15} J yr^{-1} of this being available by the year 2025.

Turning now to the generation of electricity and/or fuels from solar sources in the UK, focusing collectors of the type shown in Figure 33 are unlikely to be developed because of the small amount of *direct* winter sunlight of the type required to operate efficiently a solar furnace. Because of the progress made during the last decade in reducing the cost of solar cells, and the fact that they operate over a wide range of electromagnetic frequencies and do not require direct sunlight, there is a remote possibility that photovoltaic conversion of solar energy in the UK may become economically viable. A small specialist market already exists for applications such as telecommunication relays, beacons, boats and toys. Further reduction of solar cell costs may enhance their application at remote sites where electricity is derived from expensive sources such as diesel generators. Centralized power generation on a large scale would require a large land area, but localized generation could use roof areas of buildings. Table 6 shows that the ultimate technical potential of photovoltaic conversion in the UK is probably only around 0.09×10^{18} J yr^{-1}, requiring 100 km^2 of land area (at 10 per cent efficiency) and that only about 1×10^{15} J yr^{-1} of this is likely to be available by the year 2025.

Table 6 Solar energy prognoses for the UK[1]

Technology	Ultimate Potential J yr^{-1}	Potential for the year 2025 J yr^{-1}
Active: water heating	0.16×10^{18}	3×10^{15}
Passive: space heating	0.43×10^{18}	60×10^{15}
Photovoltaic	0.09×10^{18}	1×10^{15}
Biofuels: Dry wastes	0.45×10^{18}	220×10^{15}
Biofuels: Wet wastes	0.12×10^{18}	45×10^{15}
Biofuels: Energy crops	0.56×10^{18}	400×10^{15}
TOTALS	1.81×10^{18}	0.73×10^{18}

1. Data from the Department of Energy, Energy Technology Support Unit Report ETSU R30, May 1985.

Finally, we summarize the UK position on stored solar energy in the form of biomass (cf. Section 3.1.3). While the growth of energy crops (see below) might seem the most obvious way to generate biomass, there are numerous sources of combustable and convertable biological materials that are seldom tapped for their energy content and which are wasted. Vast quantities of *organic waste* are produced in the UK with an annual energy content approaching 10 per cent of our fossil fuel demand. These include *dry biomass* in the form of straw, wood and refuse, and *wet biomass* which includes animal slurrys, vegetable and food industry wastes. The efficient utilization of these wastes is carried out by an increasing number of local authorities; for example, the production of biogas by Birmingham Sewage Works, of steam for electricity or district heating by the GLC, Coventry, Sheffield and Nottingham authorities, and of *RDF (refuse derived fuel) pellets* at Castle Bromwich, Doncaster, Eastbourne, Grimsby, Huyton (Merseyside), Newcastle-upon-Tyne and Westbury (Wilts). Despite the recent growth of refuse conversion, however, in 1985 only 6 per cent of wastes were being recycled to create energy in the UK compared with *ca.* 75 per cent in Luxembourg and Denmark, 50 per cent in Sweden, and *ca.* 25 per cent in West Germany, Japan and France. As Sir Hermann Bondi, then Chairman of the National Environment Research Council, remarked at the conclusion of a recent biomass energy conference which discussed the wider application of such techniques: 'One could envisage industrial communities or industries which generated organic wastes using this waste to generate fuels for their own local transport purposes. Modern bureaucracies could thus make good use of their large volumes of waste paper'. A sobering thought, perhaps, and one that is under review by many hard-pressed local authorities who wish to exploit the value of waste. The ultimate potential of recycled waste in the UK to provide energy fuels compares favourably with the savings derived from other solar techniques (Table 6). In the 2025 energy scenario of ETSU, the prospects for producing biogas, methanol fluid and RDF pellets (char) are extremely attractive, contributing by that time an estimated saving of 8 million tonnes coal

(a)

(b)

Figure 35(a)   The Anglesey solar bungalow (designed and built for J B Wright) which combines the attributes of active and passive solar designs. There are 33 m² of flat-plate solar collectors on the south-facing roof. Solar-heated water from the collectors is pumped to a storage tank, thence via a heat exchanger to the domestic water supply, or direct to underfloor heating circuits. The conservatory on the left and the large patio windows provide additional solar gain during winter months. For this house, solar energy provides some 40 per cent of annual space and water heating requirements. (b) Mechanical logger as used in Oregon, USA, for converting biomass, in this case trees, into wood chips for incineration and steam raising. This particular harvesting machine can devour an entire tree in less than a minute. A tonne of wood chips generated in this way will produce the same amount of energy as 2.75 barrels of crude oil.

equivalent (*ca.* 0.2×10^{18} J yr^{-1}), more than the potential of any other 'solar' derived resource yet considered.

Even this pales into insignificance compared with the long-term prospects for biomass energy crops (Figure 35b). To obtain further increases in home-produced biomass energy, however, might necessitate changes of land use. Since present agricultural practice is aimed at maximizing *food* rather than *energy* yield, the overall efficiency of biomass production for energy in the UK is low. Therefore we would have to increase the yield of the energy crops at the expense of food production or, as this would almost certainly be considered ill-advised, to develop methods of cultivating combined food and energy crops. Some 83 per cent of UK land area is already used for crops, grazing and forestry, leaving little scope for new developments elsewhere unless *catch crops* were produced in the 6–10 week growing season after the main harvest. Other possibilities for energy crop production fall into three main categories:

(i) Forestry land at present used for timber production for buildings, paper etc. could be extended to yield energy crops, particularly if some of the 10% of the land that is mountainous and unsuitable for agriculture were cultivated. Coniferous plantations are widely used in the USA for energy crops and would be suitable because they produce biomass all year round. Another suggestion is that rapid-growing trees, not native to the UK, could grow effectively in cool temperate climates, yielding good energy crops.

(ii) Enhanced biomass energy production could also arise from the harvesting of coastal swamps, marshes and other aquatic environments where rapid growth takes place of marsh grasses, reeds, seaweeds and algae. With better management these could be cultivated and harvested as energy crops.

(iii) Finally the biomass component within existing agricultural crops might be increased without affecting food production. To do this we might improve soil nutrients, crop rotation and management, or breed genetic strains of corn and grass that produce *both* good energy and nutrient yields. Within a few decades, such techniques might reasonably be expected to double photosynthetic efficiencies in crops to about 2 per cent.

It has been estimated that with continued fuel price rises in the longer term, energy crops could become cost-effective in the UK and that bearing in mind constraints on land use, come 0.56×10^{18} J yr^{-1} could ultimately be produced, 0.4×10^{18} J yr^{-1} of this by the year 2025. Clearly from Table 6, biomass wastes and energy crops are the 'solar' schemes with the greatest immediate impact and growth potential in the UK. However, their future is fairly uncertain because of their susceptibility to both local and national political factors as well as to the economic constraints that affect other energy resources.

The ultimate potential for solar energy utilization in the UK (*ca.* 1.8×10^{18} J yr^{-1}) is a saving over 60 million tonnes of coal equivalent, about half the coal produced in the UK annually in the early 1980s. Of this about 40 per cent should be realisable within the next 30–40 years.

3.2 Wind energy

Atmospheric circulation of air masses depends on two things: (i) differences in incident solar energy over the Earth's surface, which bring about variations in air density, causing winds to blow; (ii) the gravitational and rotational energy of the Earth, which controls and modifies circulation patterns. The amount of energy involved is about 3×10^{23} J yr^{-1} (Part I, Figure 1) but, of course, only a tiny fraction of this can be tapped by windmills which are, at most, a few tens of metres high.

How do you think variations in wind speed will affect the power generated by a windmill?

The kinetic energy of the wind is proportional to the square of its speed (v):

$$\text{kinetic energy} = \tfrac{1}{2}mv^2$$

where a mass (m) of air passes a given area per second. In the case of windmill blades sweeping an area (A):

$$\text{mass per second} = \text{density of air } (\rho) \times \text{swept area } (A) \times \text{wind speed } (v)$$

The kinetic energy of the wind is partly transformed into pressure against the blades as the air is slowed down in passing through the area swept by the rotating blades. In the case of an *aerogenerator* attached to the windmill this energy is converted into an electrical output equivalent to:

$$\text{kinetic energy per second (i.e. power)} = \tfrac{1}{2}(\rho A v)v^2 = \tfrac{1}{2}\rho A v^3 \qquad (5)$$

From this cube relationship it follows that if the wind speed doubles then the power output will be increased by eight times. Of course, it is not possible to extract all the energy in the wind since to do so would mean that the air would be brought to rest. Theoretically, aerogenerators could recover about 60 per cent of the available power, but practical efficiencies about 40 per cent are more usual. Also in practice, most aerogenertors operate only within a given range of wind speeds — a certain minimum speed is needed to start the blades rotating and, at the other end of the range, the blades are progressively 'feathered' (i.e. their effective area is reduced by angling them out of the wind direction) as the wind becomes more vigorous.

A typical modern aerogenerator might have blades with an effective surface area of 350m^2. At a constant wind speed of 20m s^{-1} (45 mph) and with a generator efficiency of 55 per cent, the power output would be:

$$\tfrac{1}{2}\rho A v^3 = \tfrac{1}{2} \times 1.3 \times 350 \times 20^3 \times 0.55 \text{ J s}^{-1}$$

$$= 1.00 \times 10^6 \text{ J s}^{-1} = 1 \text{ MW}$$

where the density of air $\rho = 1.3$ kg m^{-3}

Two designs of aerogenerator are being developed, both of which range up to several megawatts output. The more traditional (Figures 36a and c) has a horizontal axis of rotation and the generator is mounted either with the rotor at the top of the tower (as shown) or at the foot of the tower, linked to the rotor by a right-angle drive and a vertical shaft. The perennial problem of keeping the rotor facing into the wind has led to the development of vertical-axis machines (Figures 36b and d) arranged around a vertical shaft which drives a ground-based generator.

Fortunately, unlike solar energy, in the UK wind movements are strongest during winter months at the time of peak demand for energy. However, wind is still unpredictable and so a storage facility is required in the form of batteries, pumped storage schemes (see Section 3.3) or via the electrolysis of water to produce hydrogen fuel (Section 5.2). Although, until the early 1980s, the UK wind energy programme was well behind those of Sweden, West Germany, and

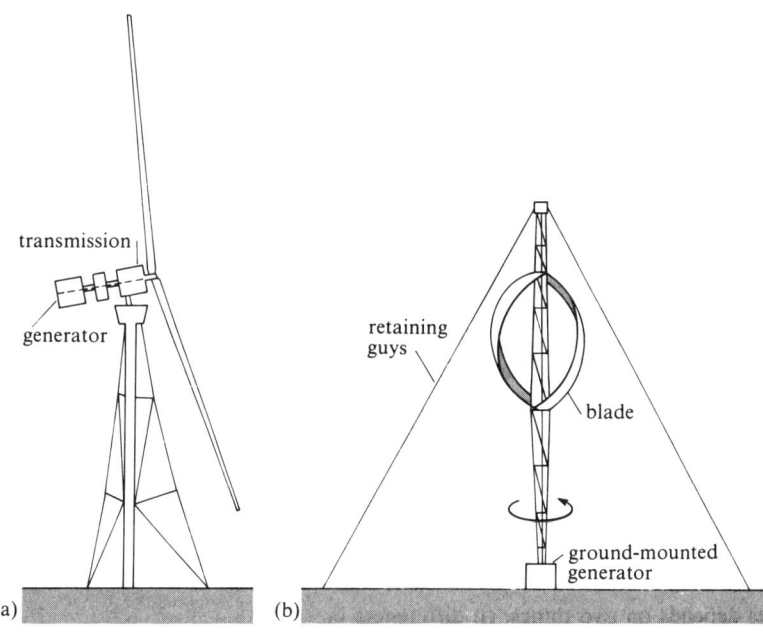

transmission

generator

retaining
guys

blade

ground-mounted
generator

(a)

(b)

(c)

(d)

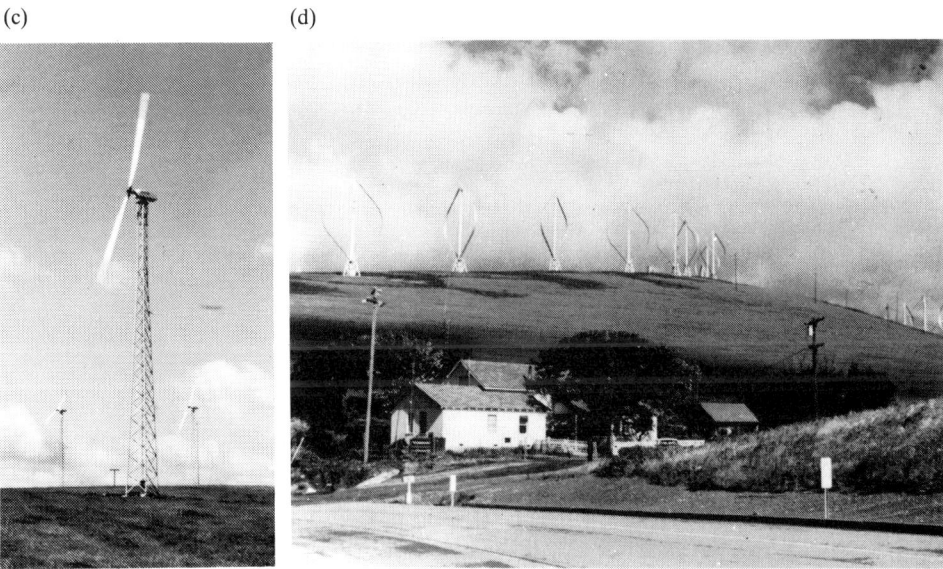

Figure 36 Wind-powered generators with (a) horizontal and (b) vertical axes (the blade twist is greatly exaggerated). (c) Horizontal axis, (d) vertical axis wind generators at the Altamont pass in California where there are several wind farms with a combined output of *ca.* 100 MW. Each windmill generates *ca.* 50–100 kW and contains a microprocessor linked to a master computer which optimises the performance of the windmill array as the speed and direction of the wind varies, and which also matches supply to demand.

particularly the USA, this is unlikely to remain the case, for wind power has become one of the front-runners among alternative energies. There are two reasons for this development: first, it has been shown that the siting of arrays of wind-powered genertors in shallow coastal waters, such as the Wash, Carmarthen Bay and Morecambe Bay, might be feasible, so avoiding the use of large areas of land. The second factor, pioneered in the USA, has been the use of advanced aerospace technology to reduce the mass and cost of large wind-powered generators.

In the early 1980s prototype machines producing 200 and 250 kW were operating in Carmarthen Bay and the Orkneys, and plans to built 3–4 MW generators during the late 1980s, including a cluster of ten in Carmarthen Bay, were well advanced. Although a long way into the future, in 1985 it was estimated that some 2×10^{18} J yr^{-1} of electrical power is technically attainable from the wind; of this some 0.1×10^{18} J yr^{-1} (3GW) might be available by the year 2025. The environmental impact will be considerable because several hundred huge wind machines spread over a large 'wind farm' area would be needed to replace just one 1000 MW coal-fired power station. Nevertheless, there are already 20 smaller scale wind farms in California and, at an estimated cost in the UK of 2.5–3.2 pence per kWh (1984 prices) for onshore schemes and 4–7p per kWh for offshore wind generation, wind power compares reasonably favourably with electricity generated from fossil fuels.

3.3 Hydroelectric power

People have used the energy of running water for many centuries; indeed, water power provided the main source of mechanical energy, chiefly through water wheels, until it was superseded by steam in the nineteenth century. Today, this renewable source of primary energy provides about 4 per cent of world energy needs: whether the energy is harnessed by a wooden wheel in a tiny stream in Nepal or by a 100 tonne dynamo at Aswan on the Nile, all hydroelectric power generation (*hydropower*) originates from the *hydrological cycle* of evaporation, precipitation and run-off, powered by the Sun's heat coupled with the Earth's gravitation.

> What is the principal advantage of flowing water over moving air as a means of power generation?

Water is very much denser than air so that, for a given speed, the kinetic energy is also much greater than for air — flowing water is a more concentrated form of energy. The principle of hydroelectric power generation is very simple: water flowing down steep gradients is channeled through pipes and gives up its kinetic energy to rotating turbines which drive electrical generators. Modern hydroelectric turbines have efficiencies exceeding 90 per cent, so most of the potential energy of water stored in a dammed reservoir can be converted into electricity.

The power developed depends on the product of the water discharge rate (Q, in m^3 s^{-1}) and the *working head* (the vertical fall of water flowing into the hydroelectric plant, H, in metres) as follows:

$$N = Kg\rho QH \tag{6}$$

where N is the power output in watts, g is the *acceleration due to gravity*, ρ is the density of water and K is the efficiency of the generating system involved. Clearly, installations that have a large working head require smaller discharge rates than do those with a small working head, as in tidal barrage schemes (Section 3.4), for equivalent power output. Thus mountainous countries, such as Norway, Switzerland and New Zealand, can rely greatly on hydropower because working heads exceeding 1000 metres are often available. However, most of the ideal sites for such installations have been used in industrialized nations and now smaller installations, using heads of 20–30 metres on broad rivers, are becoming economic (e.g. in Switzerland and Alpine France) as the cost of fossil fuel energy has risen.

Even in Britain, where the CEGB is committed to large scale plant, interest in smaller hydropower installations is increasing, often involving combined water management and power generation schemes, as at the Keilder reservoir in northeast England. The scheme is planned to produce over 6 MW from the 15 m^3 s^{-1} of water discharge. Another application of hydropower, which has been developed at several sites in Britain, is in *pumped storage schemes*, where the same water is continuously recycled between two reservoirs at different altitudes (Figure 37a).

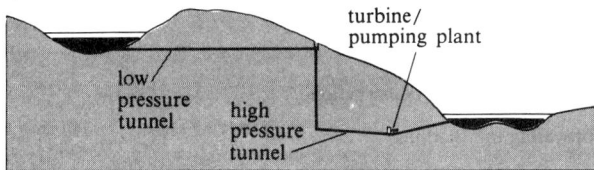

Figure 37a A cross-section of the layout of the Dinorwic pumped storage hydroelectric scheme at Llanberis, north Wales. The vertical shaft linking the low pressure and high pressure tunnels is 440 metres deep and 10 metres wide.

Low cost electricity from other power stations, usually nuclear stations which are not easily closed down over short periods, is used during the night to pump water to the top reservoir. The process is reversed to generate electricity during periods of peak demand such as occur during winter evenings (Figure 37b) especially just after popular TV programmes. The Dinorwic pumped storage scheme at Llanberis in north Wales, linked with the nearby Trawsfynydd nuclear power station, is the largest such scheme in Britain with a maximum output of 1320 MW, which can be reached in 10 seconds during an emergency. Of course, pumped storage schemes are net consumers of electricity because it takes more energy to pump water to the top reservoir than is returned when the same amount of water is released through the turbines. However, this method of *storing* electrical energy represents an enormous saving in generating costs over alternative methods of meeting peak demand, which involve operating coal-fire or oil-fired stations well below their economic capacity for most of the time.

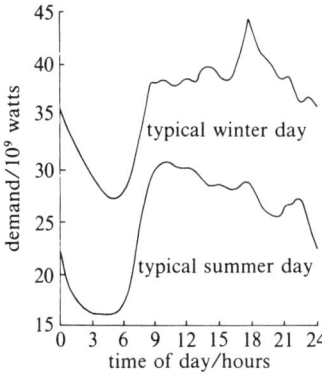

Figure 37b Summer and winter electrical power demand on the CEGB system. Note that the typical winter evening peak demand shown here is 45 GW: the highest peak demand made on the CEGB system is closer to 50 GW.

3.3.1 Contribution of hydropower to global power supplies

Figure 38 summarizes the world distribution of available and potential hydropower, a total of 2.2×10^6 MW, of which about 0.35×10^6 MW (16 per cent) was operational in the early 1980s. The world's largest hydropower station uses a 7 km long dam across the Parana river at Itaipu on the Brazil–Paraguay border. It was opened in November 1982, and by 1988 will have a generating capacity of 12 600 MW which will be difficult to absorb because the demand for electricity in Brazil has increased much more slowly than anticipated when the scheme was planned in 1970. This is yet another example that shows how political and economic factors can change radically during the long *lead times* that are required in exploiting certain physical resources.

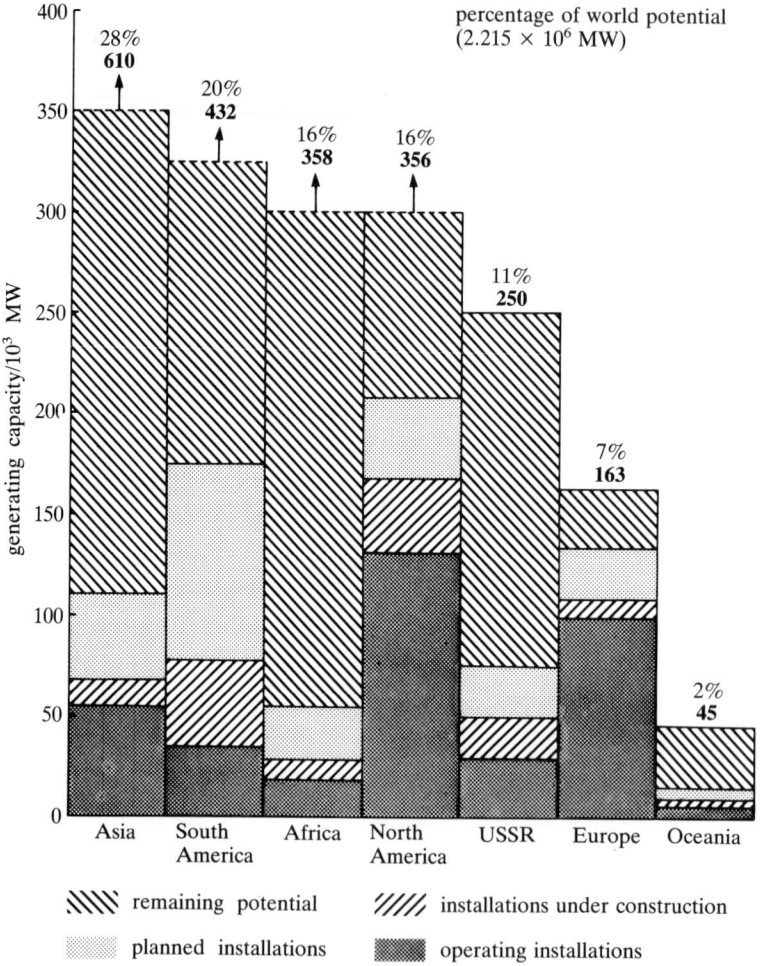

Figure 38 Stages of development of hydropower resources in different parts of the world (based on the 1976 World Energy Conference survey), showing the proportions of the total potential that are operating, under construction, or at the stage of advanced planning. The potential capacity for each area (bold numbers) is also expressed as a percentage of the capacity for the whole world.

It is clear from Figure 38 that the industrialized nations have had both the technological skills and the capital to undertake large hydroelectric schemes. Hence, the European countries have developed their hydroelectric potential most extensively, withover 60 per cent utilized, and a further 20 per cent planned or under construction. North America has also achieved 40 per cent development, and South America has the most widespread developments in the non-industrialized world with 40 per cent of its potential operating or planned. In contrast, little use is made of hydropower resources in Africa or Asia. Notice also from Figure 38 that countries with high mountain ranges and high annual rainfall are particularly favourable for hydropower development. These include the western regions of North and South America, New Zealand and Alpine Europe. Such geologically young mountain ranges are, of course, not well endowed with fossil fuels so that, in many ways, the exploitation of hydropower substitutes for conventional sources of power. In contrast, the petroleum-rich countries of the Middle East and North Africa have little hydropower potential owing to their arid climates.

Interestingly, China aims to add a further 50 000 MW to its existing hydropower production of 22 000 MW by building ten giant new hydropower centres at strategic points on its major rivers, four of them on the Yangtze. Thus China would have the most rapidly expanding hydroelectric generating industry in the world, and would produce 20 per cent of its power requirements from this source by the year 2000. To achieve this target, advanced equipment and technology will be imported, notably from France and Switzerland.

> Despite the rather optimistic view of the global contribution of hydropower portrayed so far, there are some snags involved in damming major rivers — what might these be?

First there are many conflicting claims on suitable sites, such as irrigation, river navigation, flood control or fisheries. Equally important, another major problem is finance: although industrialized nations have assisted with projects such as the Aswan Dam in Egypt and the Kariba Dam in Zimbabwe, such large schemes are frequently beyond the resources of the developing nations to maintain. Often, the capital cost must be written off over 100–200 years since major reservoirs are likely to become silted up. Both these African schemes suffered from yet another problem — the unexpected spread of disease resulting from the formation of large areas of open water where bacteria and mosquito breed. Equally as important, large hydropower schemes are often irrelevant to underdeveloped countries, which have little use for large amounts of electrical power. The main beneficiaries of large hydropower schemes in the Third World are the big industrial enterprises, such as aluminium smelters, which are commonly local outposts of multinational companies.

In view of the proposed developments in Figure 38, we can conclude that the growth of hydropower generation is likely to keep pace with, or to outpace slightly, the growth in total world energy production and demand for the foreseeable future, probably contributing between 5 and 10 per cent of those requirements. In Britain it contributes 0.5 per cent to present total energy consumption, but provides 6 per cent of electrical energy. The total installed capacity amounts to about 1500 MW, nearly all in the North of Scotland. In the UK, hydroelectric schemes have little scope for further development, except through installation of pumped storage schemes.

3.4 Tidal power

Tidal power depends on the gravitational pull of the Moon, and to a lesser extent the Sun, acting on the oceans as they rotate with the Earth. The total tidal power of the oceans amounts to 8×10^{19} J yr^{-1} (Part I, Figure 1), but there are many practical reasons why only a small fraction of this can actually be exploited. The use of tidal energy for *tidal power generation* is based on the same principles as the generation of hydroelectric power. A tidal scheme requires a large tidal basin across which a long dam (known as a *tidal barrage*) can be built to trap the water introduced at high tide. The water levels between the two sides of the barrage are balanced by controlled flow through sluices and tubines set in the barrage.

> How will the relationship $N = Kg\rho QH$ (equation 6) between power production, discharge rate and working head apply to the tidal scheme?

The head (H) will vary constantly according to the state of the tide and, although the potential discharge rate (Q) is vast, flow can occur only during those periods when there is a significant difference in water level between the two sides of the barrage. Because the head (H) available in a tidal power scheme is small, turbine efficiencies (K) tend to be much less than in hydroelectric schemes, commonly 20–25 per cent, again implying that Q must be large.

Figure 39a illustrates a simple example in which water flows through the turbine on the ebb-tide only — this is called single direction generation. The variations in sea-level outside and within the basin are shown, respectively, by heavy and light curves, and the shaded area represents the height of the head when generation can occur. Cross-sections showing water levels and flow directions are given below the curves. As you can see from Figure 39a, two periods of 4–5 hours a day only are available for power generation. The daily output of a tidal scheme can, of course, be extended by using the two-way (reversible) turbines with power generation on both the rising and falling tides (Figure 39b), known as double direction generation.

From Figure 39, it follows that a single direction generation scheme provides power for about 40 per cent of the tidal cycle whereas double direction generation increases this potential to between 50 and 55 per cent. Note that the combined periods for power output are not doubled because, in a single direction scheme, the water in the basin can be maintained at a higher average level than in a double direction scheme, and so power generation occurs for a longer period on the outgoing tide.

Although tidal power provides a diurnal output which is often out of phase with daily demand, it is a completely predictable and highly reliable source of renewable energy. This has been effectively demonstrated by the two tidal generating stations in operation today: a small experimental station near Murmansk in the USSR, which was completed in 1979, and by nearly 20 years of production from the 240 MW La Rance tidal station at St. Malo in Brittany. The main part of the La Rance dam contains twenty-four 10 MW turbines, which generate power in one direction only but which can be used in the opposite direction as pumps to increase the available head, rather like a pumped storage scheme. The minimum head at which the turbines produce maximum power output is 5.65 metres, though they can operate with heads down to 3 metres. With an average tidal range of 8–9 metres at La Rance, the equivalent of 5–6 hours a day at maximum output has been achieved, on average, since the equipment was installed (i.e. 240 MW for 5–6 hours, equivalent to 55 MW

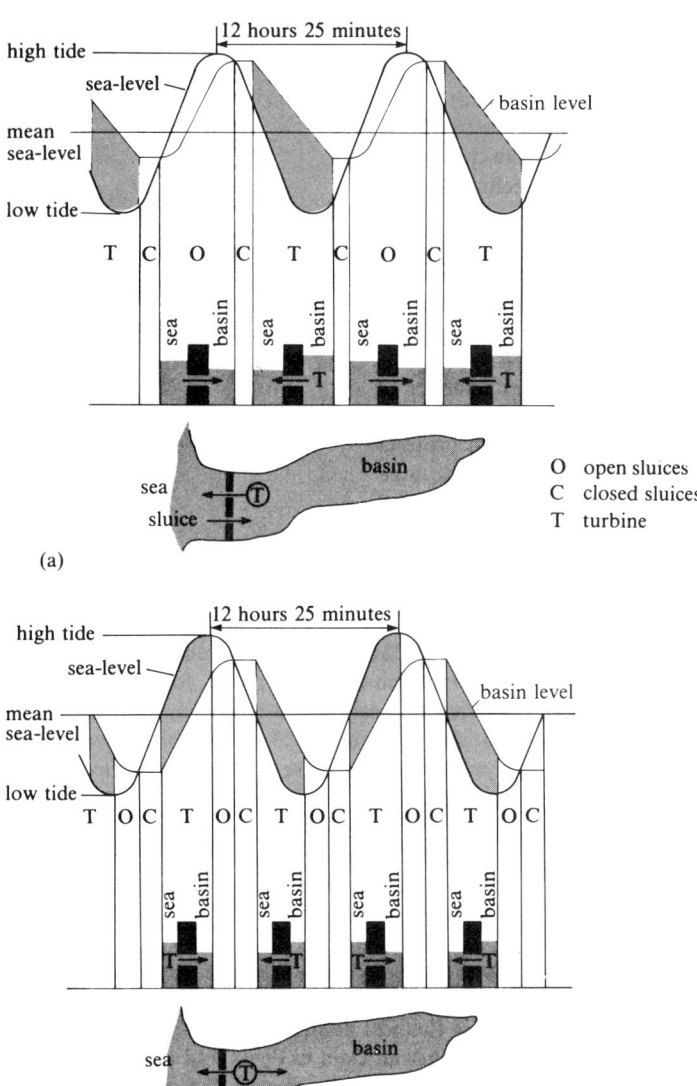

(a)

(b)

Figure 39 Diagrams illustrating the times at which tidal power generation is possible in a scheme based on a single basin. (a) Single direction generation on the ebb tide only; (b) double direction generation on both the flood and the ebb tides.

continuous power generation). The local amenities have benefited in the form of improved agricultural and water management in the surrounding countryside, a motorway across the estuary along the barrage, and improved fishing and sailing facilities. In 1985 prices, a scheme of the type at La Rance would cost £200 million, and produce electricity at about 3.8 pence per kWh, taking into account the interest and depreciation, which is only a little greater than the cost of power from conventional sources. The main environmental cost is the need for dredging operations to avoid permanent silting up of the estuary. The main advantage is that tidal energy resources are renewable.

Good sites for tidal power schemes around the world are few, mainly because of the large tidal range required. The most interesting undeveloped sites are Canada's Bay of Fundy, which has the highest tidal range in the world (17 metres) and the Severn estuary in the UK (Figure 40) where the funnelling effect of the coastline magnifies a normal tidal range of 4 metres up to an average of 11 metres between Cardiff and Weston-super-Mare. The report of the Severn Barrage Committee in 1981 showed that because of the much larger water volumes and working head than at La Rance, thirty times the power output could be guaranteed. The proposed 'inner barrage' scheme (Figure 41a) would use 160 turbines (Figure 41b) each of 45MW maximum output, a total installed capacity of 7200MW. The likely output, based on La Rance experience, would be equivalent to about 1500MW continuous (about 5×10^{16} J yr^{-1}), or 6 per cent of present *electrical* energy demand in the UK. The scheme would be kept as simple as possible with ebb-tide generation only; more complex schemes were rejected because the added costs did not lead to proportional benefits in output. However, the environmental impact of barrage construction would be widespread. Sand, gravel and hard-rock blocks would be required in enormous amounts — some 40×10^6 m^3 of aggregates for the solid embankment (Figure 41a) and similar amounts to build the 100 or more hollow concrete structures equipped with sluices and turbines, each weighing 90 000 tonnes, which would be floated into position. The downstream side of the embankment would be constructed of coarse rock fragments sufficiently heavy to withstand currents when first dumped into the sea. This barrier would be backed by an upstream extension of lower grade materials — sand, mine waste, etc., to provide mass and stability to the structure. Both surfaces, particularly

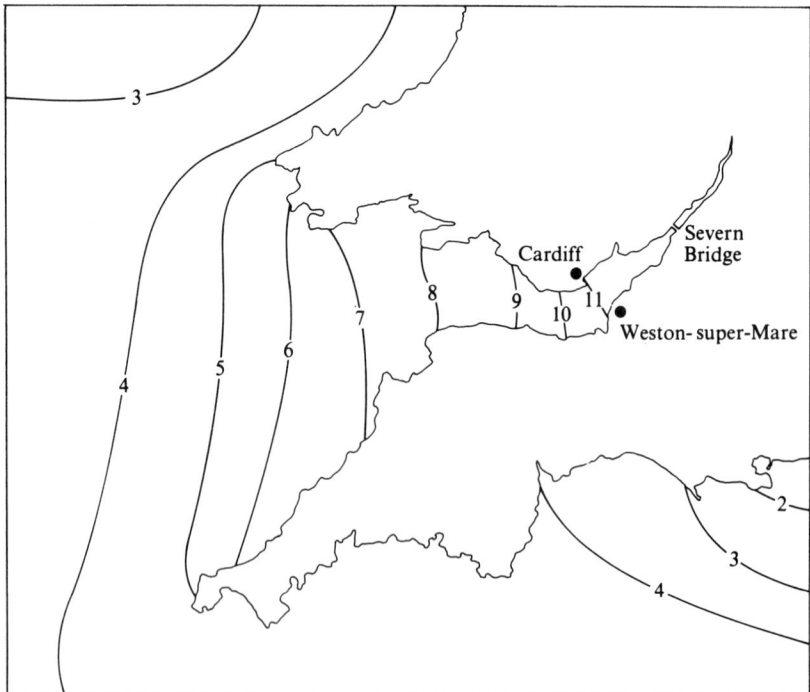

Figure 40 The average ranges (in metres) of spring tides in the Severn estuary and adjacent regions, showing how the tidal effect is amplified by the funnel-shaped channel.

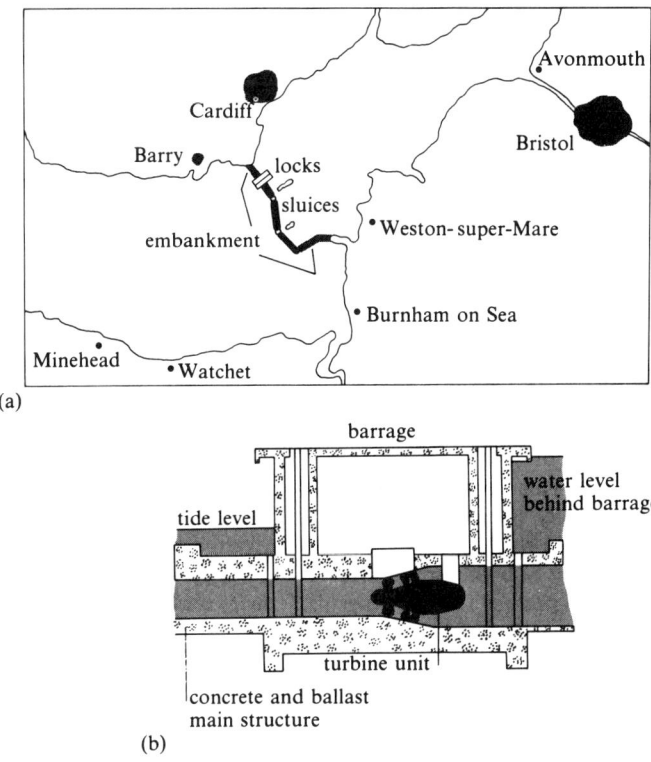

(a)

(b)

Figure 41 (a) The proposed 'inner barrage' scheme for the Severn estuary; (b) a cross-section of a turbine structure that forms part of the embankment in tidal power schemes such as that at La Rance and that proposed for use in the Severn estuary.

the seaward surface, would then be protected from erosion by a layer of coarse hard-rock blocks.

In this area, the aggregates and cement for the concrete structures would probably be obtained from the Carboniferous limestone quarries of the Mendips and South Wales, imposing a considerable environmental burden on the these areas in the form of quarrying, blasting and road transport. Much of the sand and gravel required in both the concrete structures and the embankment could come from the bed of the estuary, so causing little disruption to the surrounding country. Colliery shale from South Wales could also be used as embankment 'fill' material. Finally, it would be necessary to bring some materials, such as hardrock blocks for surfacing, from greater distances (eg. Cornish granite, Carboniferous sandstones from South Wales or the Pennines, quartzites and hard igneous rocks from the Welsh borders and Midlands).

Whereas the Severn Barrage scheme would take 10 years to build and cost £6000 million, two other interesting, smaller schemes have been proposed. The first of these is a scheme based on the 'English Stones' site, much further upstream from the proposed site of the Severn Inner Barrage, and just below the Severn bridge. This scheme is attractive because it would cost only £980 million, yet would produce *ca.* 270MW average output at only 3 pence per kWh. The main attraction of this smaller scheme is that it involves less financial risk and would provide an additional road crossing much more rapidly than the larger inner barrage. The second smaller scheme, at Strangford Lough,

Northern Ireland, would also cost less than £1000 million, would incorporate 200 MW of installed capacity and produce about 10 per cent of the province's electricity. Here, the savings in fossil fuel imports make the scheme very attractive, in addition to the social and economic benefits of the project for Northern Ireland, but neither the Strangford Lough or either of the Severn Barrage schemes had received the go-ahead by 1986. Of course, there are several less beneficial aspects of both schemes such as the problems of silting, pollution and restricted access for navigation; nevertheless, the Government and several private investors had provided funding for a detailed Severn Barrage feasibility study to take place. Preliminary studies of various social, industrial and environmental side-effects indicate a reasonable balance between those that are beneficial and marginally detrimental. A particular problem is that while our demand for electricity varies with the daylight hours (i.e. with the Sun), the times the turbines could operate most effectively vary with the phases of the Moon, so differing from day to day. Unlike the piecemeal installation of wind power generators, for example, a tidal barrage is an 'all-or-nothing' scheme and is, at present, unpopular in Government circles because it will cost the equivalent of 4 nuclear power stations to build and will generate only twice the power (calculations based on the large 'Inner Barrage' scheme). Nevertheless, there is a powerful lobby in favour because the tidal scheme would last 100 years rather than 30, would produce far less potentially damaging side-effects than those arising from nuclear waste, and would be free from the changing economics of the uranium and fossil fuel markets. Yet if the scheme were to depend solely on private investment it would appear unattractive because of the long lead time during barrage construction before any return would be forthcoming. In 1986, the dilemma between the short-term economic unattractiveness and the long-term attractiveness of the scheme for exploiting a benign power source remained to be resolved.

3.5 Wave energy

Waves are generated in the ocean by the action of winds blowing over the water, and so two main factors govern the amount of energy available from waves: the wind strength, and the uninterrupted distance they travel over the ocean, known as the 'fetch'. The geographical location of the British Isles is ideal on both counts because they lie on one of the main global wind belts and receive waves from the long fetch of the North Atlantic. The energy of a wave is translated into a wave *amplitude* — the greater the wind speed and/or fetch, the greater the amplitude (a). The total power, P, per unit length of wave front is given by:

$$P = \tfrac{1}{2}\rho g c a^2 \text{ watts per metre} \tag{7}$$

where ρ is the density of water and c is the speed of the wave.

As an illustration we may calculate the power of waves with an amplitude of 1 metre and a speed of $4\,\text{m s}^{-1}$ which is typical for waves that occur around the British coastline. Given that $\rho = 10^3 \text{ kg m}^{-3}$ and $g = 9.8 \text{ m s}^{-2}$, substituting in equation 7, we have:

$$P = 1/2 \times 10^3 \times 9.8 \times 4 \times 1^2 \text{ W m}^{-1}$$
$$= 19.6 \times 10^3 \text{ W m}^{-1} = ca. \ 20 \text{ kW m}^{-1}$$

If, say, the wave amplitude were to increase to 10 metres, as in storm conditions, then the potential power output would be increased by a factor of 100 because power output is proportional to the square of wave amplitude.

Figure 42 shows the most favoured sites for wave energy conversion around the British coastline where the power output normally exceeds that calculated above. Remember that a wave front with a power of 40 kW per metre would need to be harnessed along a 50 km length to yield the power output of a large (1 000 MW) coal-fired power station, assuming 50 per cent efficiency. This would prove very difficult for the following reasons: (i) the power density is low (50 km for 1 000 MW), so a large capital investiment is required for the converter system to produce a worthwhile output; (ii) the converter system must be built to cope with rapid short-term variations of power level, ranging up to a peak of 10 000 kW per metre, nearly three hundred times more powerful than the average; and (iii) the power output would be extremely random and unpredictable, thus requiring an energy storage device (e.g. batteries or hydrogen fuel cells — Section 5.3). Other major problems associated with the development of *wave energy converter systems* (WECS, for short) are the mechanical complexities of many designs and the need to withstand the hostile environment of the North Atlantic. Many ingeneous WECS designs have been proposed; we shall describe one that shows sufficient potential for prototype development.

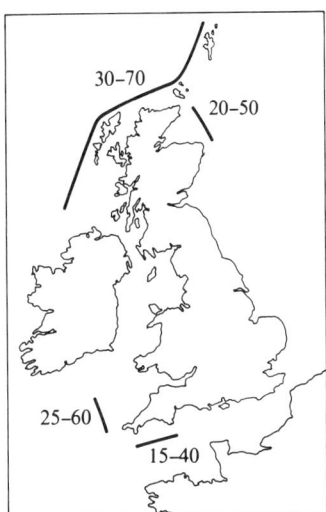

Figure 42   Optimum locations for wave energy conversion devices around the UK coastline. Data show the estimated power availability in kilowatts per metre averaged over 12 months.

Oscillating water column converters consist of a massive concrete structure with an inverted box that traps a volume of air above the surface of the ocean. As waves pass, the water level moves up and down, forcing air in and out of the box through a piping system (Figure 43). At the centre of the piping network is a four-way valve, which changes position according to the overall direction of the air flow, so ensuring that air is always forced through the turbine in the same direction.

Teams working on WECS have estimated that electricity generating costs would be 4–10 pence per kWh (1985 prices), rather high compared with conventional power stations. However, because no WECS device has yet moved beyond the scale-model prototypes being tested in laboratories and river estuaries, the final operating costs are very uncertain. The Japanese, meanwhile, have installed more than 300 small-scale versions of the oscillating water column converter to power buoys and lighthouses, each requiring 70–120 W. Their first large-scale device of this kind was installed in the Sea of Japan in 1978 and has produced 0.6 MW during trials, considerably less than the peak output of 2 MW predicted. Active interest in developing wave energy systems continues in Japan, the UK, the USA, Canada and Norway; there are plans by a UK consortium to build a prototype commercial-scale 1 MW station off the Isle of Lewis during the late 1980s.

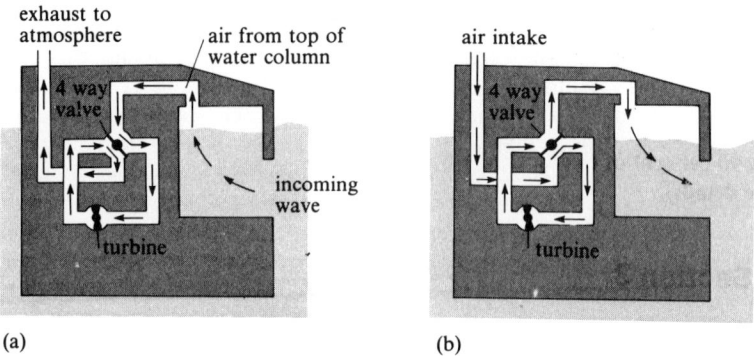

(a)   (b)

Figure 43   The operating principle of one form of the oscillating water column converter, as developed at the National Engineering Laboratory, showing details of air flow on rising (a) and falling (b) waves.

Recent estimates of the potential for wave power generation around the UK coastline suggest that some 5 GW(0.16×10^{18} J yr^{-1}) might be realized once WECS become more cost-effective than at present. This represents nearly 20 per cent of electrical energy demand in the UK in 1985. Equivalent figures for other countries involved in WECS research and development (potential and percentage of 1985 electricity consumption) are: Japan 11 GW, 15 per cent; Norway 5 GW, 50 per cent; USA 23 GW, 7 per cent.

3.6   Summary of Chapter 3

1   Surface energy resources include those derived by exploitation of solar radiation either directly, or indirectly through biomass, wind, hydroelectric and wave sources — the last two depend also on the Earth's gravitational forces. Tidal power depends only on gravitational interaction with the Sun and the Moon.

2   All these resources, except hydropower, were making little or no contribution to flobal or UK energy supply in the early 1980s, but most of them have considerable potential for power generationat economic rates.

3   The wider use of flat-plate solar collectors for water and space heating, of focusing collectors and photovoltaic solar cells to generate electricity, particularly in low latitude countries, and of biomass energy crops to produce storable and transportable fuels, could significantly reduce dependence on fossil fuels and nuclear power. Indeed, the biomass energy potential is so vast that with appropriate investment and lead time it could replace much of the demand for conventional energy resources. Table 6 provides a summary of their potential for development in the UK.

4   After early design problems, aerogenerators have now developed to cope with the rapid and fluctuating changes in power that result from changes in wind speed (power varies as wind speed cubed). Recently developed 3 MW generators could be deployed in clusters in the UK (following the example of California wind farms) on hill slopes and in shallow coastal waters, leading to an estimated maximum output of about 60 GW, or 2×10^{18} J yr^{-1} of which 3 GW might be available by the year 2025.

5   The power output of hydroelectric schemes depends on the working head and the discharge rate; therefore, such schemes are developed ideally in mountainous regions (large head) or through estuarial barrages (exploiting large tidal discharges). World hydropower production is now 1.1×10^{19} J yr^{-1} (0.35×10^6 MW, nearly 4 per cent of global energy demand); the estimated maximum potential is 7×10^{19} J yr^{-1}.

6   The hydropower potential of mountainous regions in the UK has already been realized (about 5×10^{16} J yr^{-1}), but the Severn estuary is one of the few ideal sites on a global scale for developing tidal power, with an output potential exceeding 1500 MW (5×10^{16} J yr^{-1}).

7   Harnessing of wave energy is more complex technically than tidal or hydropower conversion, requiring complex converters, some of which are being tested on a limited scale. If they were to become fully developed and deployed along 250 km of British coastline, about 5000 MW (0.16×10^{18} J yr^{-1}) of electrical power might be generated.

8   The potential in the UK of non-solar renewable surface energy resources as identified in Section 3 — wind, hydroelectric and tidal, and wave power — amount to some 2.3×10^{18} J yr^{-1} (dominated by wind power) which far exceeds present (1985) electrical energy demand (*ca.* 0.8×10^{18} J yr^{-1}). Of this only 0.15×10^{18} J yr^{-1} is realizable, according to the ETSU scenario for energy demand, by the year 2025. This compares with 0.7×10^{18} J yr^{-1} realizable from solar-based sources by this time (Table 6).

4 Side-effects of energy conversion

Study Comment. This Chapter aims to summarize and compare the long-term side-effects of the exploitation of different energy resources and thereby to evaluate the importance of these pollution risks.

Our prime concern in this book has been to explain the necessary geological and technical background on which an understanding of different energy resources and their availability can be based. Apart from a few comments on the local side-effects of the mining and conversion of energy resources, we have left aside the broader environmental issues, such as the effects on the atmosphere of burning fossil fuels, the storage of nuclear waste, and so on. It is argued by environmentally conscious social groups that much more weight should be given to the side-effects, or pollution risks, associated with the exploitation of energy resources when deciding on future energy policy. But in the early 1980s the more extreme forms of such arguments, which would mean abandoning the nuclear programme and burning less fossil fuel, had attracted little political favour in the western world.

It is increasingly important, however, for society to be aware of the various side-effects (1–11 in the list below) which are associated with use of different energy resources (A–J). An interesting exercise is for you to consider which side-effects result from the use of each resource. Our answers appear after the two lists.

A	wood-burning	1	emission of water vapour
B	coal-burning	2	emission of carbon dioxide
C	oil and petrol burning	3	emission of sulphur dioxide
D	gas-burning	4	emission of carbon monoxide
E	nuclear reactors	5	nitrogen oxide emission
F	geothermal power	6	particulate matter (dust)
G	solar power (direct uses)	7	spent radioactive fuel
H	wind generators	8	changes in water flow
J	hydroelectric and tidal	9	use of oxygen
	installations	10	lead emission
		11	unburnt hydrocarbon emission
		12	none of 1–11

The following paragraph is a summary of the most important associations in the opinion of the authors — it is not necessarily complete and many circumstances exist where other side-effects may be associated with each type of energy resource.

Wood-burning(A) : Side-effects 1, 2, 4, 6 and 9

Coal-burning (B) : Side-effects 1, 2, 3 (even smokeless fuels emit some sulphur dioxide), 4, 5, 6, 8 (power stations), 9.

Oil and petrol (C) : Side-effects 1, 2, 3, 4, 5 (especially motor car fuels), 6, 8 burning   (power stations), 9, 10, 11 (motor car fuels).

Gas-burning (D) : 1, 2, 3, 4 (rarely with natural gas), 8 (by power stations), 9.

Nuclear   (E) : 7, 8 (by power stations).
generators

Geothermal   (F) : 1, 2–5 (for volcanic gases in hyperthermal areas), 8.

Solar power   (G) : 12 (although, in power stations, 8 might become impor-
direct uses   tant).

Wind   (H) : 12 (unless you count the distrubance created by building
generators   large numbers of windmills).

Hydroelectric (J) : 8 only.
and tidal

From this list it is clear that when we tap the Earth's supply of *primary renewable energy by using solar energy or the kinetic energy of moving air and water, the side-effects are relatively few compared with those resulting from exploitation of non-renewable secondary* energy resources (fossil and nuclear fuels). Exploitation of primary resources involves only minor shifts of pre-existing equilibria: for example, silting behind hydropower dams. Contrast this with the many types of emission when coal and oil are brought into cities and burnt or with the problems of storing or disposing of radioactive wastes arising from nuclear power stations.

A further potentially important side-effect not yet mentioned is waste heat: unlike the exploitation of solar energy, where heat is extracted from and then returned to the surface environment, the thermal processing of fossil and nuclear fuels contributes to the build-up of global heat derived from stored, non-renewable energy resources. The side-effects of the exploitation of secondary resources are considered in Sections 4.1–4.3.

4.1   Side-effects of fossil fuel and biomass conversion

The exploitation of fossil fuels is still on the increase, which means that carbon dioxide, water vapour, carbon monoxide, sulphur dioxide, nitrogen oxides and particulate matter are added to·the atmosphere in increasingly large amounts.

For example, carbon dioxide and water vapour play an important part in controlling atmospheric temperatures. This is because incoming short-wavelength solar radiation passes readily through the atmosphere, but much of the outgoing long-wavelength radiation is absorbed by CO_2 and water vapour, particularly in the denser, lower atmosphere, which thus acts as a warm blanket. It follows that if there is a continued rapid increase in atmospheric CO_2 and/or water vapour, there should be more absorption of outgoing radiation, leading to a global increase in mean atmospheric temperatures.

The process described above is called the *greenhouse effect*, because of the analogous effect of glass and moist air in greenhouses, which lets in sunlight but prevents the heat escaping. Before considering the volumetric additions to *atmospheric carbon dioxide* (CO_2) and water vapour made by fossil fuels, we need to consider a few points about the natural distribution of CO_2.

What is the main source of atmospheric CO_2?

Volcanic gases from the Earth's interior are ultimately responsible for all the CO_2 in natural systems. They add an estimated 7×10^{13} kg of CO_2 a year, but since the total abundance of atmospheric CO_2 (about 2.5×10^{15} kg) is only 30–40 times greater than the annual addition, other processes must be maintaining the CO_2 equilibrium. This process is the *carbon cycle* — photosynthesis and biochemical decay circulate CO_2 in the birth–life–death cycles of the biosphere with a small proportion being preserved as fossil carbon

(Part I, Section 1.2) But this is not the only dynamic part of the carbon cycle (Figure 44): dissolution of CO_2 in surface waters and groundwaters, especially in the sea, leads to the precipitation of carbonate-rich sediments, mainly in the form of limestones. The question is whether the relatively rapid large-scale emission of CO_2 into the atmosphere from burning fossil fuels over the past 150 years could have upset the equilibria established by the carbon cycle and led to an increase in atmospheric CO_2.

In 1980, global consumption of hydrocarbon fuels was equivalent to 2.4×10^{20} J, or approximately 8×10^9 tonnes coal equivalent, containing about 75 per cent carbon, i.e. 6×10^9 tonnes of carbon were released into the atmosphere (see Figure 44). In the same way, the cumulative energy expenditure from fossil fuels before 1980 was equivalent to about 240×10^9 tonnes of coal which added 180×10^9 tonnes of carbon to the atmosphere. Of course, this carbon is immediately oxidized on combustion and is, therefore, equivalent to $180 \times 10^9 \times 44/12 = 660 \times 10^9$ tonnes of carbon dioxide (CO_2). The total mass of the atmosphere is 5×10^{15} tonnes, so the CO_2 liberated prior to 1980 is equivalent to:

$$\frac{660 \times 10^9}{5 \times 10^{15}} = 1.32 \times 10^{-4} = 132 \text{ parts per million}$$

Now, in the mid-nineteenth century, before large-scale use of fossil fuels began, the equilibrium abundance of atmospheric CO_2 was 460 p.p.m., but it has since been increasing, for reasons which are now obvious: it follows that the 1980 level of atmospheric CO_2 would have been $460 + 132 = 592$ p.p.m. if all that released from fossil fuels had stayed in the atmosphere.

Following this calculation it is, perhaps, surprising to learn that the observed abundance of CO_2 in the atmosphere in 1980 was only 520 p.p.m., so only about half the CO_2 added by fossil fuel combustion has contributed to a build-up of CO_2 in the atmosphere. One possible explanation is increased biological productivity, increasing the vegetation cover, so removing atmospheric CO_2. But vegetation cover has been reduced by growing populations and related urbanization, so this is hardly likely to be the right answer. Most of the extra CO_2 liberated by fossil fuel combustion has probably gone into the sea, which is an enormous reservoir, quite capable of acting at least as a partial buffer, accommodating increased amounts of CO_2 by increased carbonate precipitation. The rate of limestone formation must have increased slightly over the past 100 years!

Figure 44 shows that the major volcanogenic CO_2 input to the atmosphere (*left*) is passed via equilibrium reactions (*right*) into carbonate sediments, and it is now clear that about half the input from fossil fuel combustion (3×10^9 tonnes as carbon) follows the same path, the other half remaining in the atmosphere. Calculation of the climatic effect of continued increases in atmospheric CO_2 are subject to many uncertainties, but suggest a rise of surface tempertures of between 1 and 4.5°C over the next 100 years. The worst result, $+ 4.5$°C, is based on a near-doubling of CO_2 abundance to 920 p.p.m. — a most unlikely ocurrence — which would cause the melting of the ice caps and the drowning of low-lying coastal areas. Thus, the thermal consequences of CO_2 emissions are not yet regarded as particularly serious.

The effect of adding to *atmospheric water vapour* (H_2O) by fossil fuel combustion is even more problematic. In the first place, such additions may be cancelled out by the hydrological cycle, which responds to increased vaporization by promoting condensation and rainfall. But if CO_2 emissions were to cause a rise in temperature, then the atmosphere might hold more water vapour, so enhancing the greenhouse effect. Equally, of course, the corresponding increase of cloud cover might cause cooling so, once again, no certain predictions can be made. Whatever changes in CO_2 and H_2O levels result from fossil fuel combustion, however, the conversion of biomass energy crops at *replacement levels* would have no effect since the CO_2 and H_2O removed by photosynthesis would be returned to the atmosphere at the same rate. However, like combustion of fossil fuels, biomass conversion does contribute to the build-up of particulate matter, or dust, in the atmosphere.

After emission, fine particles (less than 2 μm in size) of ash, smoke and soot may remain in the atmosphere for many years; collectively these are known as *atmospheric dust*. Smoke and soot particles consist of oil droplets and carbonaceous residues, often impregnated with tar, and result from incomplete combustion of fuels. The dust particles can act as net *absorbers* of solar radiation, so causing the atmosphere to be heated, or they may serve as nuclei for the formation of water droplets, thus adding to the cover of *reflecting clouds*, and causing a *cooling* effect. Although again complex, the net effect is one of cooling. For example, fine volcanic dust eruptions from Mount St. Helens in 1980 reduced average temperatures by 0.04°C in the northern hemisphere and by 0.1°C in the north polar area. Indeed, quantitatively, natural emissions of dust exceed those from fossil fuel burning by about an order of magnitude.

The substances considered so far have a potential effect on climate but are otherwise not harmful. But several other substances released from fossil fuels have toxic properties if released in sufficient concentrations, so their effect is most noticeable in urban areas.

Carbon monoxide is released by incomplete combustion of fossil fuels, especially from motor vehicles. On entering the bloodstream it reduces the blood's capacity to carry oxygen, and adverse effects on vision and time-sense result from exposure to 200 p.p.m. for 15 minutes, or to 50 p.p.m. for about 2 hours. The air of city streets commonly has 50 p.p.m., whilst levels of several hundred parts per million can occur in traffic tunnels. Fortunately, the 200 million tonnes of this gas being released into the atmosphere each year (Table 7) do not accumulate but are probably fixed into carbohydrates by photosynthesis.

Sulphur dioxide is emitted when coal and, to a lesser extent, oil are burnt, because both fuels contain sulphur or sulphides; since vast amounts of these fossil fuels are being consumed, *atmospheric sulphur dioxide* (SO_2) is becoming

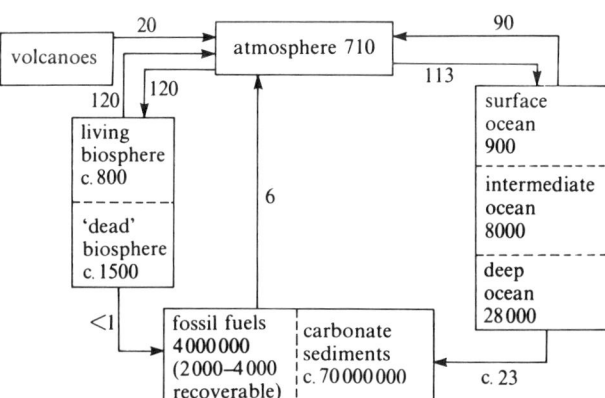

Figure 44 A smiplified representation of the principal natural carbon reservoirs and fluxes (reservoir data in 10^{12} kg; fluxes in 10^{12} kg yr^{-1} of carbon).

a major global pollution problem. To the natural, mainly volcanic supply of 140 million tonnes a year, fossil fuel combustion and metal sulphide smelting add 75 million tonnes of sulphur every year (Table 7). Although removal mechanisms are fairly rapid, these include washing out by rain and snow:

$$2SO_2(gas) + O_2(gas) = 2SO_3(gas)$$

sulphur   atmospheric   sulphur
dioxide   oxygen   trioxide

$$SO_3(gas) + H_2O(liquid) = H_2SO_4(liquid)$$

sulphur   rainwater   sulphuric
trioxide   acid

Table 7   Summary of the amounts (tonnes per year) of various components added to the Earth's atmosphere from natural sources and from human activities

	Natural	Human
carbon dioxide	7.3×10^{10}	2.2×10^{10}
water	5.0×10^{14}	1.0×10^{10}
carbon monoxide	?	2.0×10^{8}
sulphur	1.4×10^{8}	7.5×10^{7}
nitrogen	1.4×10^{9}	1.5×10^{7}

This sulphuric acid takes the form of fine mists, or *aerosols*, which are highly irritating to animal respiratory systems, have a major corrosive effect on buildings and, through their geochemical effects on soils, poison the root systems of trees and other vegetation. Rainwater, with pH values as low as 4.0 (so-called 'acid rain'), falling in parts of Scandinavia, West Germany and Poland, has caused severe damage to forests and fish life. From Figure 45, which gives contours of the total sulphur deposition over western Europe, it is clear that the highest values coincide with enhanced levels of emissions related to power stations and factory chimneys of industrialized areas.

A significant factor that has actually reduced the level of SO_2 emissions in Britain over the past 30 years was the implementation of the Clean Air Act in 1956, which has progressively reduced domestic coal burning, a major source of low-altitude atmospheric pollution. Other factors have been the replacement of steam by diesel and electric locomotives and of 'town gas' by natural gas (cf. Part I, Section 7.5). New technologies for desulphurizing coal and petroleum are now being developed as, too, are low-cost *catalytic processes* for removing sulphur from chimney gases. So there is every reason to believe that high-altitude atmospheric SO_2 emissions from industrial chimneys and ground emissions from vehicles can also be substantially reduced.

Nitrogen oxides, N_2O, NO and NO_2, are formed by the oxidation of nitrogen during the combustion of all fossil fuels, particularly at the high temperatures found in internal combustion engines and power station boilers. Although the natural turnover of nitrogen from biological and agricultural sources is high (Table 7), much of this is in non-polluting forms, and the emissions of pollutant nitrogen (*ca.* 1.5×10^7 tonnes a year) are quantitatively much less than those of sulphur. However, *atmospheric nitrogen oxides* are a problem in some industrial areas, particularly in sunny areas, since sunlight completes the oxidation process of N_2O and NO to NO_2; the aerosols produced have an acidic and irritating effect similar to that of SO_2 aerosols. They are also the chief constituent of

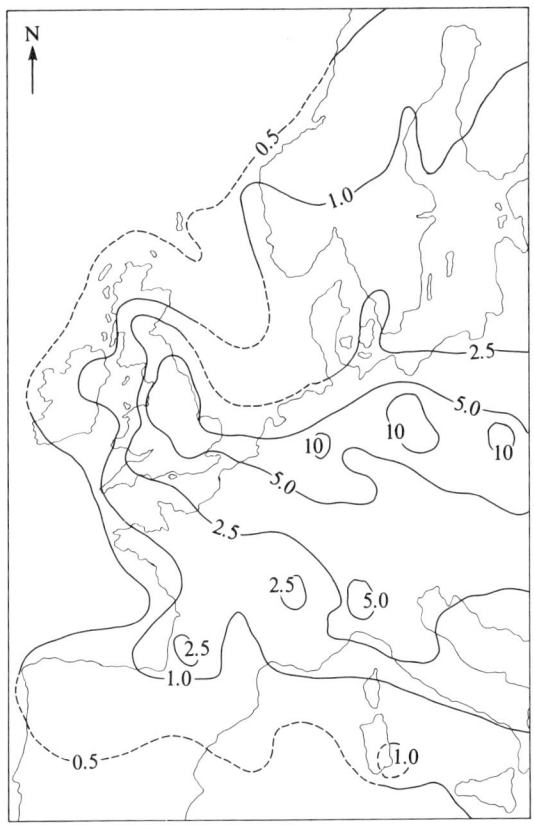

Figure 45 The pattern of deposition of sulphur over north-west Europe and Scandinavia for the year 1974 (units are gm^{-2}).

'photochemical smogs', for which Los Angeles is particularly well known because it lies in the fog-prone basin of sunny California and is well populated with motor cars. This is the main reason why there has been substantial interest indeveloping methanol-fuelled vehicles, which are less polluting, for use in California (cf. Section 3.1.3).

Lead pollution arises because organic lead complexes are added to refined petroleum to improve engine performance and act as anti-knock agents. Because lead is a poison that is alleged to cause brain deterioration, especially among children, a vigorous debate has ensued about whether the lead in petrol should be removed or reduced. In the USA, lead-free petrol is used in most vehicles manufactured since the mid-1970s; in the UK the lead content of petrol is being reduced form 0.4gl^{-1} to 0.15gl^{-1} as a first step to using lead-free petrol by the end of the decade. Several other EEC countries have adopted the same policy; West Germany, in particular, has announced its intention to have lead-free petrol in the near future.

Fossil fuels also contain significant amounts of other trace elements, which are emitted during combustion and which have attracted some public concern; for example, arsenic, cadmium, mercury and uranium (see Section 1.2.6).

4.2   Side-effects of the nuclear power industry

Nuclear power generation involves concentrated fissionable fuels which, after fission, leave large volumes of product isotopes, some of which are highly radioactive. Much of the criticism levelled against the industry falls under four main headings:

1 The effect of abnormal radiation levels on health through damage to cells causing malignant cancers, genetic defects, etc.

2 The operational safety of the industry whereby dangerous emissions could arise from reactor operations (e.g. Three Mile Island in 1979 and Chernobyl in 1986) and from the transport of irradiated fuels to or from the reprocessing plant (cf. Section 1.1.2).

3 Disposal of radioactive waste.

4 The increased potential for the proliferation of nuclear weapons: as more nuclear stations are built and the traffic in spent and reprocessed fuels increases, so do the chances of plutonium being hijacked for weapons.

As you know from Section 1.3, since about 1975 these four side-effects have been partly responsible for the decreasing growth rate of the industry, particularly in the USA, though to a much lesser extent extent elsewhere (e.g. western Europe and the USSR). The strongest arguments against further expansion of the nuclear power industry are those of operational safety and waste disposal, so, assuming that the industry does expand, strict security measures, improved fail-safe devices and basic research into waste disposal techniques are of paramount importance. The remainder of this Section is devoted to the latter topic.

4.2.1 Radioactive waste disposal

Most products of nuclear reactors are solid at ordinary temperatures. They cluster around atomic mass numbers 90 and 140 (equations 2 and 3, Section 1.1) and among the common nuclides (with *half-lives* in brackets) are: ^{90}Sr (28 years), ^{96}Zr (10^{17} years), ^{131}I (8 days), ^{135}Cs (2×10^6 years), ^{137}Cs (28 days) and ^{142}Ce (5×10^{15} years). Besides these fission products, a range of actinide nuclides, accompanying the plutonium-238 extracted during reprocessing, also occur in lesser amounts. Decay curves, expressed in terms of heat output, are shown in Figure 46 for the fission products and the actinide nuclides from a 1 GW AGR; clearly, there is a big drop in the emission of heat from radioactive fission products after 600 years. But it is agreed that it is inadvisable for much of this waste to enter the biosphere for the next 10^3, and preferably the next 10^4–10^5 10^4–10^5 years.

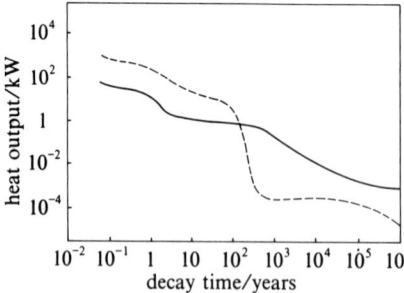

Figure 46 The heat output from the actinides (solid line) and radioactive fission products (dashed line) resulting from the spent fuel elements of a 1 GW advanced gas-cooled reactor.

Before considering disposal methods, we need to know the scale of the problem. By referring back to Figure 2, we can determine the mass of waste products (ie. radioactive nuclides *other* than those of uranium and plutonium) which must be disposed of during the 30 year lifetimes of different reactors. For a 1000 MW burner reactor, assuming that 50 tonnes of plutonium are produced, then 4800 – (4600 + 50) = 150 tonnes of waste products arise. For a 1000 MW breeder reactor, the equivalent mass is only 55–(20 + 6) = 29 tonnes of waste products. If these waste products were to be in solid form with a density of 10,000 kg m^{-3}, then the volumes of waste involved would be 15 m^3 for the burner reactor and 2.9 m^3 for the breeder reactor.

So the volume of waste from each reactor is small — the space occupied by a few large wardrobes. But for some of this waste, well shielded storage space is required. The wastes produced by the nuclear power industry, together with those from other medical and scientific practices, are sub-divided for disposal purposes into those with low, intermediate and high levels of radioactivity.

Low-level wastes include: *gases* with short half-lives, a few years at most (e.g. ^{41}Ar, ^{3}H and ^{222}Rn), which are vented to the atmosphere; *liquids*, which are discharged into the sea or rivers; and *solids*, which are burnt in incinerators, buried on purpose-built tips, or packed into drums to be tipped in the deeper parts of the oceans.

Intermediate and high-level wastes pose the common problem that almost all that has been produced during the history of the nuclear power industry is, at present, stockpiled pending agreement on disposal sites and methods. All the high-level wastes produced in the UK are stockpiled in liquid form at Sellafield in double-skinned stainless steel tanks surrounded by concrete and cooled by sets of stainless steel coils through which water is circulated. About 5 m^3 of high-level liquid waste is produced per tonne of fuel reprocessed. This volume of fluid is 50 times greater than the solid volume if it has a density of 10000 kg m^{-3}, so various methods are being developed to solidify these wastes in the form of synthetic rocks and minerals or as borosilicate glass, which could be incorporated in a thermodynamically stable multiple containment system (Figure 47) ready for disposal. The next question is, where should these containers be dumped?

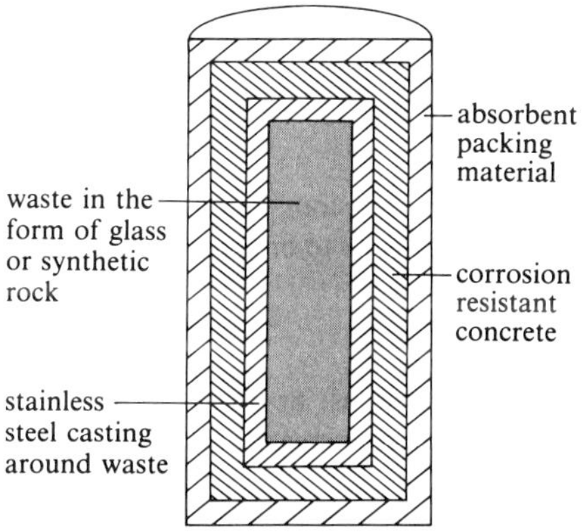

Figure 47 Schematic representation of a cylindrical multiple containment system that would be suitable for burial of nuclear waste in a deep borehole.

In the best circumstances, it is predicted that the containment system will last for 100–1000 years and that the glass itself will inhibit the migration of radioactive nuclides for a further 1000 years. So, in view of the long decay times (Figure 46), the ideal geological site for disposal should also act as a barrier. On land, the prime geological contenders for waste containment in the UK are clay-rich rocks, salt deposits and hard, crystalline, igneous or metamorphic rocks.

What properties do all these rock types have in common that will be useful in impeding the migration of radioactive wastes?

They are all relatively impermeable to fluid migration, and have large-scale homogeneity and a massive nature. Salt and some clay deposits have the advantage of being self-sealing, owing to their high plasticity; clays have a high capacity for ion absorption, and crystalline rocks can be identified which are generally dry, stable against geological movement, extensive and inert to any heating effect from the waste canisters.

Looking at the problem in a UK context we will consider these three groups of rocks in turn, viz. clay-rich sedimentary rocks, salt deposits and crystalline rocks. Clay-rich sedimentary rocks occur in the Mesozoic — Tertiary basins of Hampshire, London and the Yorkshire-Lincolnshire areas, and among the Palaeozoic rocks of central Wales, the Lake District and south Scotland. The latter might provide the better disposal sites as the older rocks are more compact and stable, and have undergone low grade metamorphism to slates. Appropriate salt deposits are of Permo-Triassic age and these underlie the Midlands, particularly the Cheshire area, Dorset and, of course, the North Sea. Hard Palaeozoic and Precambrian crystalline rocks are generally found in the upland reigons of Wales, Northern England and North Scotland.

Clearly, there is no shortage of suitable rocks in the UK. Figure 48a shows the kind of underground operation that is envisaged for disposal, consisting of repository areas 300–1000 metres below ground with the waste containers placed in boreholes spaced along the floor of a series of horizontal tunnels. In the early 1980s, the UK government put into abeyance its plans for the investigation of disposal sites, pending the outcome of similar work in the USA and France. But as some wastes have already been awaiting disposal for 20–30 years the problem will become increasingly in need of a solution and there are signs (in 1986) of renewed interest in test-drilling at selected sites.

Meanwhile, research into disposing of wastes in the sea-bed continues (Figure 48b). Three techniques are being studied: (i) containers dropped on the surface of the ocean floor; (ii) projectiles that penetrate the soft sediments of the shallow ocean crust; and (iii) disposal of waste canisters into boreholes drilled deep into consolidated ocean crust. Of these, the second looks potentially most interesting and cost effective; in geological terms, once radioactive nuclides escape the containment system, it is anticipated that migration rates through ocean-floor sediments would be very slow, no more than 1 metre per 10000 years. Moreover, it has been pointed out that the local opposition to waste disposal would be less prevalent than on land, there being no voters in the ocean deeps.

Unfortunately, the ultimate disposal of intermediate and high-level nuclear waste, whether below the continents or the sea-bed, is unlikely to be far advanced before the turn of the century. By that time, Britain will have accumulated $2000 \, m^3$ of solid waste and $4500 \, m^3$ of liquid waste and, of course, there will be a growing number of sealed-off radioactive mounds following the decommissioning of present-day nuclear power stations. Whatever the future of the nuclear power industry, one thing is clear: high-level wastes are here to stay for the next 10^3–10^4 years at least.

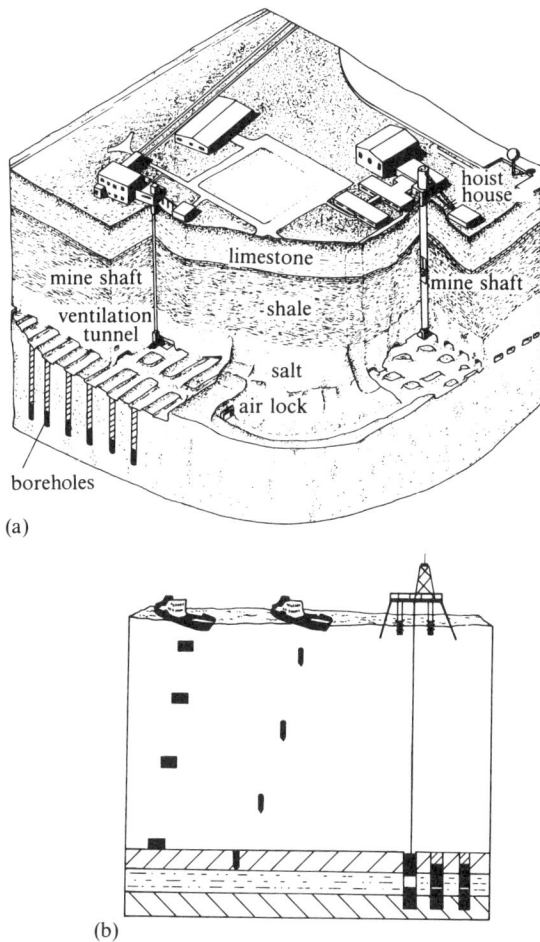

(a)

(b)

Figure 48 (a) Simplified cut-away model of land-based repository for the disposal of nuclear waste based on the proposed salt repository at Lyons, Kansas, USA. The waste containers would occupy the lowest quarter of the boreholes, which would then be filled with clay. (b) Alternative methods for the disposal of high-level nuclear wastes on or below the sea bed.

4.3 Excess heat

Over recent geological time the Earth's surface has been in thermal equilibrium: mean surface temperatures in the atmosphere and hydrosphere have remained essentially constant. At present, 85–90 per cent of our global power production is derived from non-renewable sources that are not part of this equilibrium, namely fossil fuels or nuclear fuels. Either directly, or indirectly after serving some useful purpose, this power is released into the atmosphere as waste, or *excess heat*; in other words, it is going towards raising the ambient surface temperature of the Earth.

By comparison with the solar input (about 10^{24} J yr^{-1}; Part I, Figure 1), excess heat (about 2.6×10^{20} J yr^{-1} from non-renewable energy resources) appears to be insignificant. But the massive *local* use of energy in urban and industrial centres turns these centres into 'heat islands' — an effect enhanced by roads and buildings, which are good absorbers of solar radiation. Thus, for example, the average summer temperature in central Scotland, the north-west and west midlands of England, and in the London area is 1 °C higher than in surrounding regions. The same effect occurs in river, lake and coastal waters that are used to cool the condensation units of power stations. Nevertheless, although these effects are small, even beneficial at present, more serious local problems involving the build-up of excess heat could occur if the use of non-renewable energy resources increases by a further order of magnitude.

4.4 Summary of Chapter 4

1 The exploitation of renewable energy resources leads to relatively few harmful side-effects compared with those resulting from (i) gaseous and particulate emissions following fossil fuel conversion and (ii) radioactive waste disposal and other problems associated with nuclear power generation.

2 Acid rain and photochemical smogs, which arise through the absorption of SO_2 and NO_2 by atmospheric water vapour, are the most worrying side-effects of fossil fuel conversion and are already polluting biological systems within industrialized areas. There is also concern that continued increases in the levels of atmospheric CO_2 and H_2O and in particulate emissions could cause a noticeable effect on global climatic patterns during the next 50–100 years.

3 The side-effects of the nuclear industry can be viewed on two scales. First there is the slow increase in the volumes of intermediate and high-level radioactive waste for which safe disposal methods are not yet agreed — hence the increased risk of local leakages into the environment. Second, there are worries about the widespread immediate and devastating effects that could result from a major reactor accident, or the non-peaceful uses of plutonium.

4 The continued release of energy from non-renewable resources into the atmosphere as excess heat must raise the global mean temperature, but this effect is, at present, slight compared with local changes.

5   The future of energy supply and demand

Study Comment. We start this final Chapter by summarizing the data on energy use with particular reference to the UK, and conclude with a brief look at possible developments, including nuclear fusion, in future energy supply.

5.1   Energy supply, end-uses and conservation in the UK

Figure 49 illustrates the pattern of *energy flow in the UK* . You will need to spend a few minutes analysing this figure. It is now becoming a little dated, but is the most complete analysis available at the time of writing. By the mid-1980s the equivalent primary energy fuel consumption figures were: coal 3.1, down from 3.47; oil 3.4, down from 3.94; natural gas 1.9, up from 1.17; nuclear 0.4, up from 0.24; hydroelectricity 0.06, up from 0.05. Thus from the mid-1970s to the mid-1980s there has been some small shift away from coal and oil towards natural gas and nuclear fuels. Bear this in mind as you read on; otherwise, the essential principles portrayed by Figure 49 remain unchanged. First identify the four principal sectors:

1   *Total production* — coal from the mine, oil and gas from the well, etc.

2   *Primary fuels** — that proportion of the total production that goes to serve the UK economy after losses in the production industry are allowed for.

3   *Secondary production* — the output from oil refineries, power stations, coke ovens, etc., but including some primary fuels (e.g. coal and natural gas) despatched to users.

4   *Final, or end-use energy* — the aggregate energy that is consumed by the individual users.

Table 8 summarizes the data in Figure 49 for secondary energy products; make sure that you can identify on Figure 49 the source of each line of data in the table. Note that, in terms of conversion efficiencies, natural gas requires virtually no treatment, so losses are negligible and this is one reason for the growth in natural gas utilization over the last decade. Similarly, oil refineries are highly efficient owing to the small amount of process heat required. In the coke ovens that process most solid fuels, the volatile content of the coal is lost in carbonization, part of the gases being burnt to heat the coke ovens, and conversion efficiency is down to 84 per cent. However, the conversion efficiency of power stations is very low owing to the low inherent efficiency of the steam turbine cycle. In fact, steam turbines operated at just over 100 °C are a little less than 33 per cent efficient — we have not compensated for the small hydroelectric contribution to the generating industry which has been included with the sources in Table 8. Furthermore, an average of 17 per cent of electrical power production is lost in transmission before it reaches the consumer, so that overall only about 27 per cent of the energy input is put to a useful end.

The net overall conversion efficiency in the UK from total production (9.26 × 10^{18} J yr^{-1}) to end-use (6.43 × 10^{18} J yr^{-1}), as shown in Figure 49, is 69 per cent — a value which has varied only by 1–2 per cent since the mid-1970s data on which Figure 49 is based. Note that this will be reduced still further by the

---

* In this context, 'primary' means the fuels first produced from each energy resource, and does *not* refer to the origin of the resource itself (cf. Part I, Chapter 1).

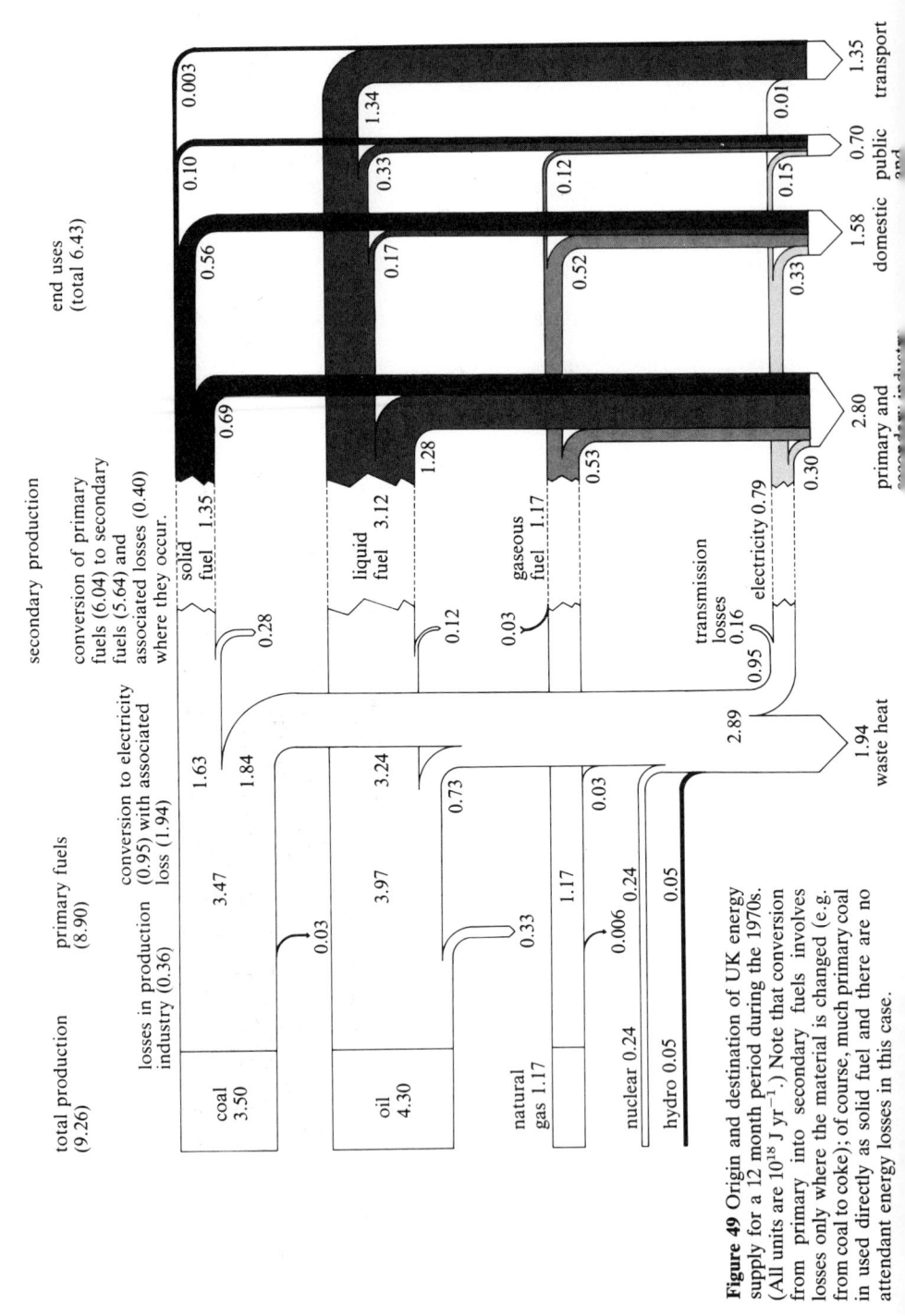

Figure 49 Origin and destination of UK energy supply for a 12 month period during the 1970s. (All units are 10^{18} J yr^{-1}.) Note that conversion from primary into secondary fuels involves losses only where the material is changed (e.g. from coal to coke); of course, much primary coal in used directly as solid fuel and there are no attendant energy losses in this case.

inefficiency of certain end-users; e.g. domestic coal fires are even less efficient than power stations, and internal combustion engines convert only 25 per cent of the energy content of their liquid fuel into motive power.

Table 8 Secondary energy production in the UK (data from Figure 49)

Secondary energy products	Energy flow/10^{18} J yr^{-1}		Conversion efficiency (%)
	input	output	
solid fuels (coke, etc.)	1.63	1.35 (+0.02 as gas)	84
liquid fuels (refineries)	3.24	3.12 (+0.01 as gas)	97
gaseous fuels	1.14	1.14	100
electricity production	2.89	0.95	33
electricity transmission	0.95	0.79	83

Clearly, then, to make available a certain amount of end-use energy in a convenient form, a much larger amount of energy resource production is required, some of which is used in the conversion process. The worst offender is the steam turbine electricity generator which, owing to the thermodynamic laws governing its operation, cannot achieve efficiencies much greater than 30 per cent. It follows that the greatest amount of waste heat arises from electricity production and that this represents nearly 70 per cent of the energy content of the resources used in the generating plant; 60 per cent of electricity production goes into 'essential', non-heat uses such as lighting, motive power, electronics and electrochemistry. The remainder is used for space and water heating, and this is the area of energy use where there is the greatest scope for substituting other forms (e.g. natural gas, thermal collection of solar energy) as these are non-essential end-uses of electricity.

Yet comparisons over the 20 years from 1960 to 1980 show that an increasing proportion of the country's delivered energy is in the form of electricity — from 6.5 per cent in 1960 to 12 per cent in 1980 — mainly because of its convenience. Clearly, in view of the unavoidably low efficiency of steam turbine generators, the proportion of total energy resource production used to generate electricity has increased by much more than this factor. Hence, the future contribution of electricity to end-use energy is important in determining future resource needs. In 1978, the total UK electricity generating capacity was 67.2 GW (2.2×10^{18} J yr^{-1} if run continuously) with another 15.8 GW of plant under construction. The hierarchy of plant being used for diffrent periods of time is shown in Figure 50. Generating plant is installed to meet peak-load demand which has, in fact, remained close to 50 GW; the base-load has been supplied by relatively new and reliable nuclear plant (Figure 50).

A particularly important factor which has the potential to limit the use of electricity and reduce the demand on basic resources as a whole is the increasingly important subject of *energy conservation*. Most conservation measures involve relatively small end-use devices or adaptions that reduce the amount of energy needed for a particular task. Home insulation and improved furnace designs are obvious examples of ways in which heat energy can be saved, whilst conservation in the transport area can arise from the wider use of public transport, car pooling, improved maintenance of vehicle engines and reduced speed limits. In particular, it is worth noting that heat losses from a typical house can be cut by almost an order of magnitude by incorporating cavity wall insulation, fiberglass loft insulation and double glazing. It is hardly surprising, therefore, that the statutory insulation requirements for new buildings have been doubled in the UK since 1975 and that grants are now available to assist with the cost of insulating older properties and public buildings.

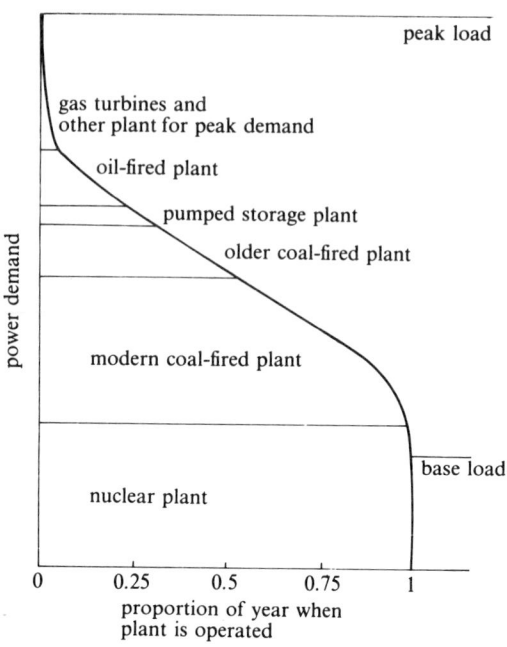

Figure 50 A graph to show how much during a year the different types of generating plant are used in the UK, and the order in which they are used as the power demand rises. The base load is the minimum continuous demand.

Another way of conserving energy is to improve the efficiency of the heat source, and this brings us back to the low efficiency of electricity generating turbines. In order to extract the maximum generating capacity from steam turbines, the lowest practical exit temperature has to be achieved, resulting in vast volumes of cooling water at about 30°C, too cool for any use apart from heating greenhouses. Overall thermal efficiency can be improved, however, by designing power stations to produce both electricity and useable heat; these are known as *combined heat and power, or CHP stations* (Figure 51). In such stations a small part of the generating efficiency is sacrificed so that hot water, and in some designs steam, can be drawn off at usefully high temperatures for industrial, horticultural or local domestic heating schemes of the type described for semi-thermal geothermal resources in Section 2.2. In this way, the overall efficiency of the CHP scheme can approach 70 per cent.

Of course, there are some difficulties; for example, power stations need to be smaller than those built hitherto, which produce far more excess heat than can be used locally. Moreover, CHP schemes need to be sited in urban locations with high levels of population, so that there is a demand for both heat and electricity. These factors, combined with the unpredictable demand for space heating load in Britain, explain why CHP schemes have not become more widely developed. Nevertheless, 15 per cent of industrial electricity requirements are met by such systems, usually where power generation is linked to heat produced by industrial processes, as in parts of the paper and chemicals industries. A small contribution to domestic heating arises through district heating schemes, the earliest of which was based on the Battersea Power Station and Pimlico housing estates in London.

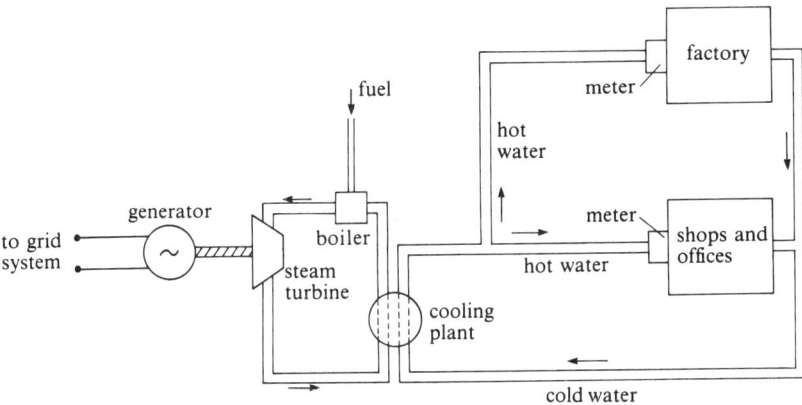

Figure 51 Schematic diagram of a combined heat and power station showing heat loads and power output.

In the sense that the energy extracted from them represents a saving in other energy resources, waste-derived fuels (disucssed in Section 3.1.4) are relevant to energy conservation schemes. So far, such schemes have not been widely developed because of their high running costs, small output and inflexibility compared with conventional, centrally co-ordinated CEGB generating plant.

Looking to the future, over the next 20–30 years it is forecasted that demand for primary energy resources in the UK will increase at rates between 0.5 per cent (low growth scenario) and 1.5 per cent (high growth scenario) to reach a level between 10×10^{18} and 13×10^{18} J yr^{-1} (compare with 8.9×10^{18} J yr^{-1} in Figure 49) by the year 2010. In the electricity sector, some 40 GW of new capacity will be installed by 2010 which, bearing in mind the loss of outdated power stations, is planned to bring peak-load capacity to about 90 GW. A shift away from the use of oil for electricity generation accounts for the slight drop in UK oil consumption since the mid-1970s (see earlier) and, by 1983, 75 per cent of UK electricity generation was coal based, with 18 per cent arising from nuclear stations. However, an important factor in the electricity supply industry's capability to survive the potential crippling coal miners strike of 1984 was its ability to switch 50 per cent of present power stations from coal to oil supply. Such flexibility is seen as an essential element in the programme for future power station building, a programme in which the nuclear contribution is likely to assume greater importance (on present policies) during the foreseeable future. The preferred technology for electrical energy supply after 2020 is the fast breeder reactor and, at the time of writing, very considerable efforts were being made to restore public confidence in nuclear power. Transboundary electrical energy flow is also seen as advantageous and a 2000 MW capacity link is now being installed between the UK and France across the English Channel.

So far as the total supply of primary energy is concerned, some 60 per cent of UK demand is, at present, derived from crude oil and natural gas which were developed rapidly during the 1970s to the point where the UK is more than self-sufficient in energy with substantial earnings from exports. No matter how much additional investment arises through tax and other incentives, there is likely to be a steady decline in North Sea production after 1990. The probable trend for primary energy supply will therefore include some increases in the use of coal and nuclear power, the latter perhaps increasing by 5–10 times over the next 50 years, with decreases in oil and gas consumption. On this scenario, the electrical component of end-use energy would increase, a trend which may be resisted because of the inflexibility of electricity compared with other energy resources which can be traded and stored more easily. Whatever the future holds for UK energy supply, however, there is no question that substantial changes in the pattern of resource consumption are inevitable. In particular, any short-term fluctuations, such as increased oil consumption due to price reductions in 1986, are likely to be smoothed out by changes in demand; the long-term prospects for dwindling crude oil reserves remain unchanged.

5.2 The future world energy scene — prospects and possibilities

The UK is fortunate in having large enough coal reserves to produce most of our electrical power — an important factor that has delayed the need to expand rapidly the generation of power from nuclear or renewable resources. Other western European countries and Japan are faced with much starker choices either of upgrading their nuclear generating programme (as has happened in France) or of increasing their dependence on imported coal and oil whilst developing renewable energy supplies (e.g. Japan). Figure 52 shows a recent scenario covering world energy supplies past, present and future, which we shall use in considering whether each of the individual energy resources is likely to satisfy projected demand.

Coal Even with the most optimistic forecasts for the development of nuclear and renewable energy resources, coal has a vitally important part to play in the world's energy future. A recent World Coal Study (WOCOL) project involving sixteen nations recognized the probability that 1977 production (3.4×10^9 tonnes a year world-wide) will need to be doubled by the early years of the twenty-first century. It is highly unlikely that coal production in western Europe could be expanded at such a rate. Although much of the world's coal resources are in the USSR and China, it is possible that four countries — USA, Australia, Canada and South Africa — could meet the needs of the western world provided that the industry could be expanded rapidly. The situation in developing countries is quite different. With the notable exception of India, few have any significant indigenous coal industries though about 50 have coal resources, so a large expansion in coal production and use in developing countries is projected in the WOCOL study.

Oil and natural gas During the 1960s and 1970s, world consumption of oil and gas grew twice as fast as that of all other energy resources. Whilst there may be temporary surpluses at times of industrial recession, there is no prospect that this rate of increase in consumption could continue indefinitely (see Part I, Chapter 8) even if oil shales were to be commercially developed. Despite the (probable) temporary reduction of world oil prices in 1986, the dwindling prospect for any major and long-lasting increase in the supply of oil and gas at acceptable prices constitutes the main reason for the increased importance of other energy resources in Figure 52b.

Nuclear energy The expansion of the nuclear industry suggested in Figure 52b exceeds even that projected in the OECD high growth curves in Figure 24, which would put severe pressure on uranium resources, though perhaps revitalizing a depressed industry. However, delays in the commissioning of new nuclear plant in some countries, combined with public reaction against the industry (Section 1.3) mean that the nuclear contribution to the growth scenario in Figure 52b is unlikely to be realized.

Renewable energy resources At present, hydroelectric power is the only surface energy resource making a substantial contribution to global energy supplies, but there is considerable potential for expansion of various renewable resources, particularly in the developing countries. Geothermal resources are already providing useful amounts of energy in some areas, and additional resources are being developed. Solar energy, both direct conversion and via biomass conversion, probably has the greatest long-term potential of the renewable resources for exceeding global production levels of 10^{20} J yr^{-1} (as required in Figure 52b). The development of other surface energy resources may have local impact (e.g. wave and tidal power in the UK) but is unlikely to provide a global solution to energy problems. However, there is one source of energy which, so far, we have not considered in this book, and which could solve all our energy problems if it were to become available, and that is fusion power.

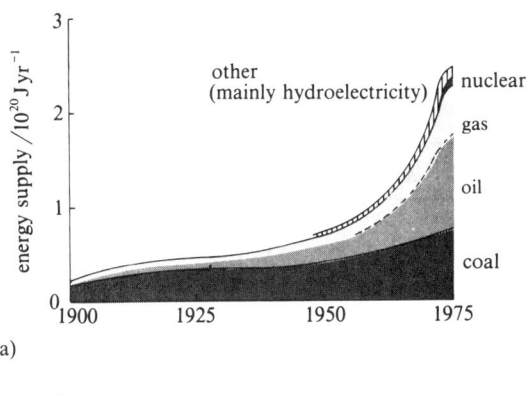

(a)

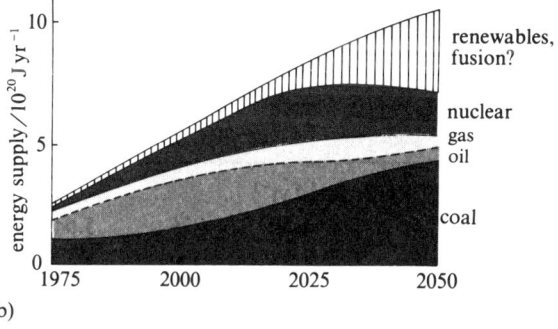

(b)

Figure 52 (a) UK Department of Energy Technology Support Unit data for world energy supplies from 1900 to 1975; (b) a scenario for world energy demand during the period 1975–2050. The latter assumes a fairly low growth rate, 2 per cent a year, and is said to be 'coupled to a cautiously optimistic supply pattern for fossil fuels and thermal nuclear energy' — see text for a discussion of how realistic this scenario might be.

Nuclear fusion power   The basis of nuclear fusion is collisions between nuclei of light atoms at very high velocities to produce large nuclei with a mass less than the aggregate mass of the small nuclei. The surplus mass is converted into energy according to the Einstein relation, $E = mc^2$. Uncontrolled nuclear fusion is the basis of the hydrogen bomb and the source of all solar energy, but can it be harnessed in a controlled way as an energy resource?

The fusion reactions which have been studied are those of deuterium and tritium — the heavy isotopes of hydrogen — and of lithium, e.g.

$$\underset{\text{deuterium}}{{}_{1}^{2}\text{H}} + {}_{1}^{2}\text{H} \rightarrow \underset{\text{helium}}{{}_{2}^{3}\text{He}} + {}_{0}^{1}\text{n} + 5.4 \times 10^{-13} \text{ J} \tag{8}$$

Deuterium comprises 0.015 per cent of the hydrogen atoms in natural waters, so it is abundant in the oceans. They contain 4.2×10^{13} tonnes of deuterium which, if extracted, could produce 3.4×10^{30} J of energy when fused, a factor of 10^7 greater than the *fossil fuel bank*. The main problems in sustaining fusion reactions are (i) that nuclei must be brought to within 10^{-15} metre of each other before the strong nuclear attractive forces can overcome electrostatic repulsions, and (ii) that to achieve the necessary kinetic energies temperatures of about 10^8°C are required. The maintenance of very high temperatures and the containment of the fuel at appropriately high pressures in fusion reactors are opposing requirements which have defeated many attempts to harness this energy resource. Nevertheless, if the technological problems are solved — and best estimates say that this will be in the early decades of the twenty-first century at the earliest — then energy future scenarios could look very different from that in Figure 52b.

If nuclear fusion or other alternative energy resources were to become widely developed in place of conventional hydrocarbon resources, this would lead to the question: what can be used to fuel smaller self-contained power units? The convenience of the petrol engine has resulted in its proliferation across the face of the Earth. Clearly, a transportable fuel is needed and, again, hydrogen could come to the fore as both a substitute for petrol and a means of storing energy until it is needed.

Outside the sophistication of a fusion reactor, the use of hydrogen as a fuel depends quite simply on the reaction:

$$2\text{H}_2 + \text{O}_2 = 2\text{H}_2\text{O} \tag{9}$$

which takes place when hydrogen is burnt in air and is exothermic, yielding 1.21×10^8 J kg^{-1}; this compares very favourably with the calorific value of petroleum products (0.5×10^8 J kg^{-1} for petrol). There are just two problems — the density of liquid hydrogen is ten times less than that of petrol, and hydrogen boils at −253°C (20K) — so there are major, but not insoluble, difficulties of transport and storage.

Hydrogen can also be used as a storage medium to smooth out the vagaries of unpredictable fluctuations in energy supply from renewable sources. Figure 53 illustrates the storage concept; in this case wind energy is used to electrolyse water into its component gases — the reverse of reaction 9 and so an endothermic process. The energy used is thus 'stored' in the gases and can be released when they are recombined. Burning the hydrogen is one way of doing this; another way is to feed hydrogen to the anode and oxygen to the cathode of a *fuel cell*, where they are recombined into water. As a result of ionization

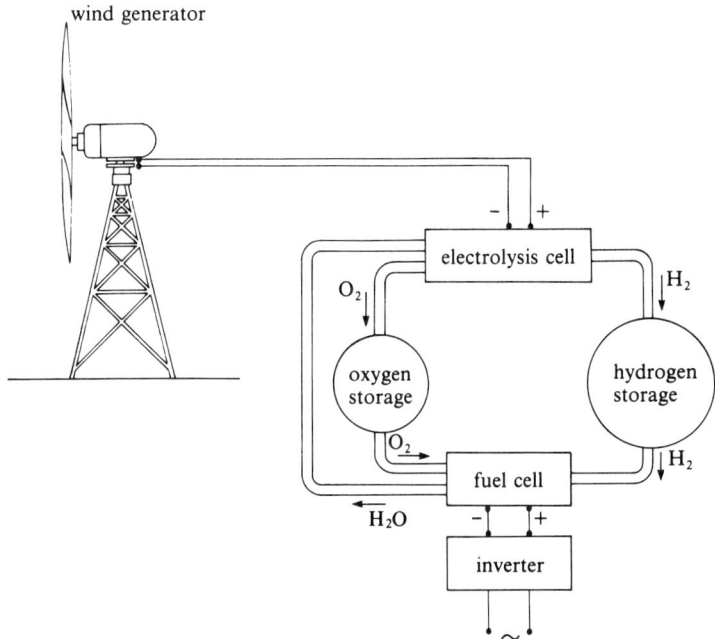

wind generator

Figure 53 Storage of unpredictable energy from wind-powered generators may be achieved by electrolysis of water into its component gases followed by their recombination in a fuel cell at times when electrical energy is required.

processes there is a transport of ions between the electrodes of the cell and a flow of electrons in the external circuit, producing a useable direct electic current. The inverter shown in Figure 53 is required to convert this to alternating current for use on an electricity supply gird.

Electrolysis is just one of several methods by which hydrogen for fuel cells could be produced; other processes involve the thermochemical splitting of water by high temperature chemical reactions and the biochemical liberation of hydrogen by certain kinds of plants and algae during photosynthesis. All three processes form part of a futuristic integrated system which has been suggested (Figure 54) for energy management based on hydrogen as the energy carrier, and known as the *hydrogen economy*. Rather than being liquified, gaseous hydrogen would be stored in high pressure vessels or, in the longer term, perhaps in depleted aquifers or oil wells. Whether nuclear fusion is harnessed or not, the hydrogen economy offers a promising alternative to the present fossil fuel economy.

With the plans for nuclear fusion and the hydrogen economy we have come a long way from the energy scenario in Figure 52b which assumes, perhaps incorrectly, that these plans will not be developed before 2025. We conclude this book with a final down-to-earth summary of energy resource futures. First, there is no question that a time will soon come when oil is restricted to a much smaller range of premium uses than in 1986. Although the long-term contributions of fossil fuels to the scenario in Figure 52b might be realized, it is less likely that the nuclear industry can be expanded at the forecast rate, and the increased contribution of renewables beyond the year 2025 could only arise realistically if fusion were harnessed. It follows that we may be forced to revise downwards the energy demand curve in coming years. One of the most

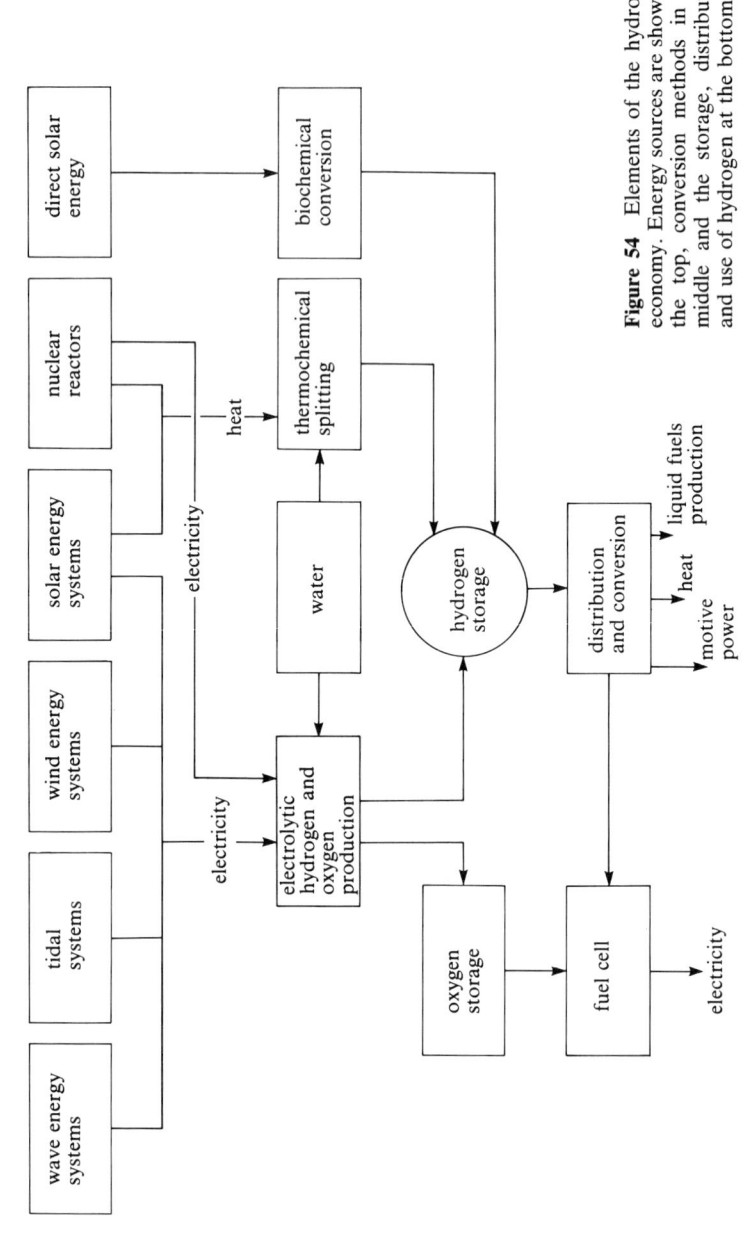

Figure 54 Elements of the hydrogen economy. Energy sources are shown at the top, conversion methods in the middle and the storage, distribution and use of hydrogen at the bottom.

fascinating paradoxes in physical resource studies of recent decades is that the rising price of oil during the 1970s, which signalled the need to begin a transition away from oil, has, in fact, reduced the demand for energy and made the need for change less urgent. Together with political factors, this is the reason for the instability of oil prices at the time of writing. Inevitably, however, change must come, but the decreased urgency with which decisions are being made may cause the western world, one day, to look back and see OPEC as a blessing in disguise.

5.3 Summary of Chapter 5

1   The overall conversion efficiency from total energy resource production to end-use energy in the UK is almost 70 per cent, but that for electrical power generation, which is contributing an increasing proportion of end-use energy, is only about 30 per cent.

2   Appropriate conservation schemes could include better standards of building design and insulation, more economic production and use of low-temperature heat and improved efficiency of heat sources, particularly through CHP and district heating schemes.

3   Most scenarios for the future exploitation of energy resources, either in the UK or on a world scale, recognize that supplies of oil and natural gas are limited and will need to be conserved. This will involve the increased use of combinations of coal, nuclear and renewable energy resources on a scale to be determined by the extent to which energy demand grows. Major production increases in any of these three resource areas would require positive political decisions coupled with financial investment.

4   Nuclear fusion, involving reactions at exceedingly high temperatures and pressures could, by exploiting the deuterium content of the oceans, theoretically produce far more energy than any other resource. However, several decades of development work, at least, are required before even prototype reactors are available.

5   Another futuristic possibility is that of energy storage based on the recycling of hydrogen fuel between electrolysis cells (energy input) and fuel cells (energy output). This is the hydrogen economy; the energy output could equally be adapted for motive power provided that the problem posed by the low density of liquid hydrogen can be overcome.

Further Reading for Part II

1. The Open University, 1984, S238, *The Earth's Physical Resources*. O.U. Press.

2. Uranium: Resources, production and demand, 1983. Joint report of the OECD Nuclear Energy Agency and the International Atomic Energy Agency, OECD, Paris.

3. P R Simpson, J A Plant and G C Brown (eds.), 1982. *Uranium 81*. The Mineralogical Society, London.

4. Vein-type and similar uranium depostis in rocks younger than Proterozoic, 1982. International Atomic Energy Authority, Vienna.

5. The Geothermal Resource, 1979. Petroleum Information Corporation, (A C Neilsen Company), Denver, Colorado.

6. Nuclear Energy in Britain, 1981. Pamphlet No. 28, HMSO, London.

7. Prospects for the Exploitation of the Renewable Energy Technologies in the United Kingdom, 1985. Energy Technology Support Unit Publ. R30, AERE Harwell.

8. D Boyles, 1984. Bio-energy, technology, thermodynamics and costs. Ellis Horwood, Chichester.

9. L A W Bedford, 1983. Wind Energy — a progress report. Coal and Energy Quarterly. 37, 19–261.

10. Tidal Power from the The Severn Estuary, 1981. Department of Energy Paper No. 46, HMSO, London.

11. R Shaw, 1982. Wave energy — a design challenge. Ellis Horwood, Chichester.

12. I Smith, 1978. Carbon dioxide and the 'greenhouse effect' — an unresolved problem. IEA Coal Research, Report ICTIS/ER 01, London.

13. K D B Johnson, 1979. The disposal of high-level radioactive wastes, *in* The British Association Meeting 1979 paper 'Energy in the Balance', Butterworth, London.

14. D Crabbe and R McBride, 1978. *The World Energy Book*, Kogan Page.

Self-assessment Questions for Part I

SAQ 1 (Chapter 1) Decide, giving reasons, whether each of the following statements is true or false.

(a) Although winds and waves are secondary effects of differential solar heating combined with the Earth's rotation, the power they produce is still regarded in resource terms as primary.

(b) Tidal energy depends on the gravitational attraction between the Sun and ocean water which raises bulges that travel around the Earth as it spins.

(c) The heat flowing through the Earth's surface, which is produced mainly by the decay of long-lived radioactive isotopes, is approximately equivalent in total energy terms to the solar energy that penetrates the atmosphere and reaches the surface.

(d) The surface temperature of the Earth is maintained at about 15 °C because the atmosphere is relatively transparent to incoming short-wavelength solar radiation, but traps much of the outgoing long-wavelength terrestrial radiation.

(e) The isotope ^{235}U which, today, is regarded as the most important source of nuclear energy, is the least abundant of the four important long-lived radioactive isotopes in the Earth.

(f) The fossil fuel bank represents all the fossil hydrocarbons that have escaped combustion and have become trapped in sedimentary formations during the last 400 Ma.

SAQ 2 (Chapter 2) Identify and summarize (in a few sentences each) the three developments in the history of the exploitation of energy resources that have led to the greatest changes in the pattern of consumption and relative use of different resources.

SAQ 3 (Chapter 2) With reference to the information in Chapter 2.2 and 2.3, match each of the statements and predictions A–E with one of the following groups of energy resources 1–5.

1 Coal
2 Oil
3 Natural gas
4 Nuclear fuels
5 Renewables (including hydroelectric power)

A   Important sources occur in deposits of most geological ages.
B   Predicted to show the lowest growth rate in exploitation between 1980 and 2010.
C   Major sources of highest quality material in Permian and Carboniferous strata.
D   Includes the greatest contribution from new technologies by the year 2010.
E   Predicted to be produced at about 0.9×10^{20} J yr^{-1} in the year 2000.

SAQ 4 (Chapter 3) There are literally hundreds of cycles of sedimentation in a coalfield like that of the East Midlands, with possibly over 100 coal seams. However, only a few beds containing marine fossils have been identified, but these can be widely recognized throughout the British coalfields. How can this be explained? How can these marine bands aid a geologist in interpreting the sequence?

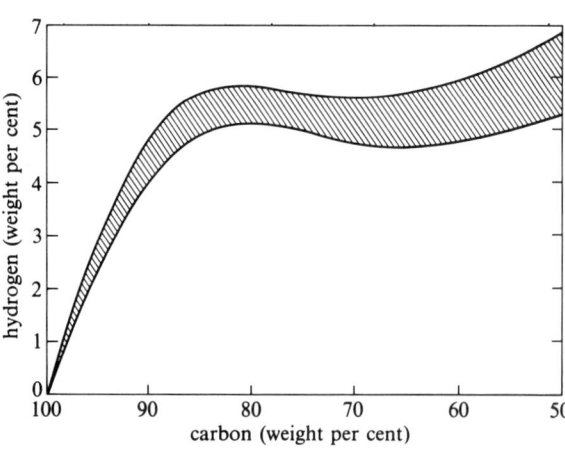

Figure 55 A graph illustrating relationship between carbon content hydrogen content over the rank ser

SAQ 5 (Chapter 3)   Figure 55 illustrates the relationship between carbon and hydrogen contents over the rank series. Mark the boundaries of the six rank stages and describe the physical characteristics of each stage.

SAQ 6 (Chapter 4)   (a) Study the cross-section of the midlands coalfields from Cheshire to Lincolnshire coalfields in Figure 17 and describe the sequence of geological events that occurred in the evolution of this coalfield. (b) What geological limits can be drawn to the area in which concealed coalfields may be located in southern England? (Figure 15 may assist you.)

SAQ 7 (Chapter 4)   Explain the meaning of the term 'stripping ratio' in relation to open-cast mining. Why can this ratio be greater for anthracite coals than for coal used in power stations?

SAQ 8 (Chapter 4)   When a coalface first enters a new area of reserves it may encounter a variety of geological hazards, which may halt it. Which hazards may be described as (a) gradual changes, (b) sudden changes? What would be the effect on production of each of the two classes of hazards?

SAQ 9 (Chapter 4)   Why is the coal produced from the modern mechanized coalface well suited to the needs of the power stations but not so appropriate for domestic customers?

SAQ 10 (Chapter 5)   (a) Using the data in Table 6, determine what percentage of the world's coal production comes from the countries of the southern hemisphere.

(b) What percentage does Australia contribute to the southern hemisphere total?

(c) Could these percentages have been predicted from the distribution of recoverable reserves given in Table 7?

SAQ 11 (Chapter 6)   Figure 56 shows the relationship between the atomic ratios H:C and O:C for petroleum at the different stages of maturation.

Area C represents natural gas, the areas labelled B refer to crude oil and the three A areas represent the three types of kerogen. Which of the A areas corresponds to which type of kerogen? (Use Table 10.)

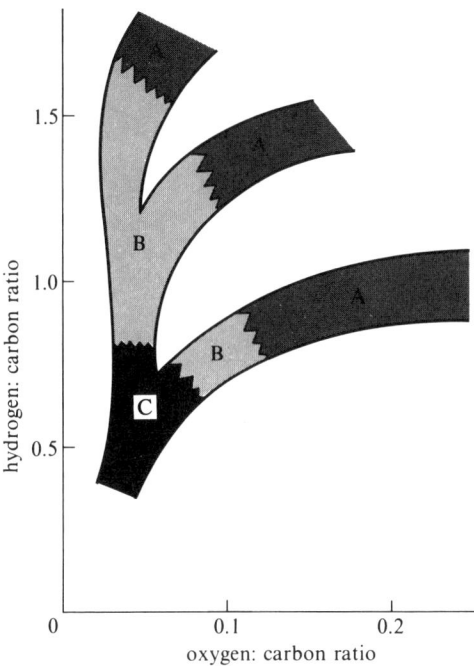

Figure 56 A diagram showing the hydrogen-to-carbon ratios and oxygen to-carbon ratios in the three types of kerogen.

SAQ 12 (Chapter 7) Examine Figure 57. Assume that the sandstone reservoir was folded prior to the migration of petroleum out of the source rocks, and also assume that the oil and gas separated into two distinct layers in trap 1.

Earth movements then tilted the whole area downwards towards the west, and as result each of the three traps contains only one of the following: oil, water or gas.

(a) What would you predict each trap should contain?

(b) What kind of petroleum deposit could form where the reservoir is open to the surface?

SAQ 13 (Chapter 7) In Figure 58 at which of the points 1 to 7 would you expect petroleum to accumulate in the reservoir? (Assume that the faulting has not hindered migration of oil and has not permitted oil to escape to the surface along the faults.) Where would you look for an oil seepage or a solid petroleum deposit?

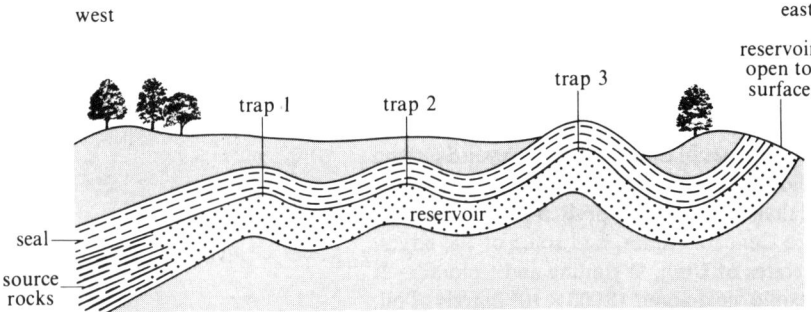

Figure 57 A series of traps in a sandstone reservoir.

surface

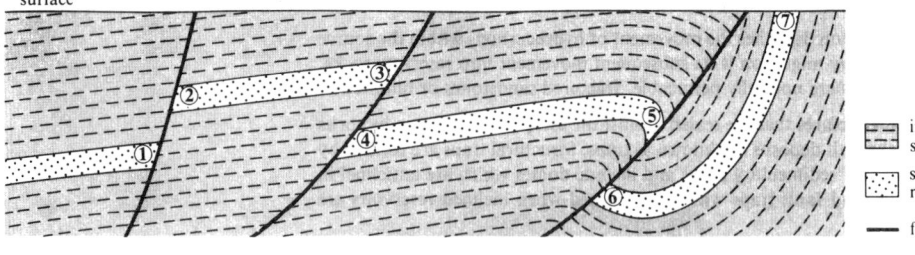

impermeable
shale

sandstone
reservoir

faults

Figure 58

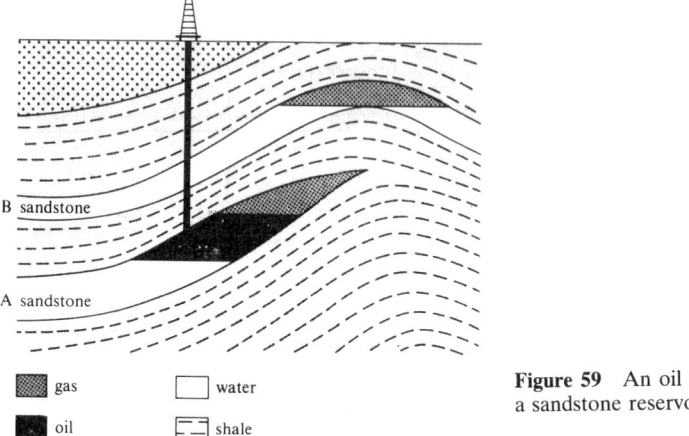

B sandstone

A sandstone

gas   water

oil   shale

Figure 59 An oil well penetrating a sandstone reservoir.

SAQ 14 (Chapter 8)   List the techniques that could be used to find an offshore oil trap and to test if there is oil within it, and the methods used to determine the shape of the trap and reservoir.

SAQ 15 (Chapter 8)   What data would you require in order to produce an estimate of reserves of oil remaining in an oil field that had been in production for 5 years?

SAQ 16 (Chapter 8) Study Figure 59.

(a) An oil well has penetrated the oil-bearing sandstone A, and with falling production from the field it has been decided to use secondary recovery methods to improve recovery. Where would you site four boreholes to achieve this? (Two would be used to obtain water and gas, and the other two to pump water and gas down into reservoir A.)

(b) What kinds of trap are depicted in Figure 59?

SAQ 17 (Chapter 9)   (a) What is the main reservoir lithology in the North Sea?

(b)   Major dry gas reserves have been found in the southern North Sea — what is the source of these gas deposits and what is their age and that of the reservoir rocks?

(c)   The major oil deposits are located in the northern North sea. What is the likely source rock for these deposits. What is the age of these sources rocks of the Forties field?

SAQ 18 (Chapter 9)   Why has the Middle East got such enormous petroleum reserves?

SAQ 19 (Chapter 9)   Assuming that all the world's oilfields have been discovered, what two main factors will decide how long the reserves will last?

Self-assessment Questions for Part II

SAQ 1 (Chapter 1)   (a) In burner reactors, a moderator is used to slow down the neutrons and the position of the control rods is continually adjusted to absorb excess neutrons. Why is it necessary to slow down neutrons and to absorb the excess?

(b) All uranium-based reactors breed plutonium but some do so faster than others: why is this?

SAQ 2 (Chapter 1)   Decide, giving reasons, whether each of the following statements about uranium ore deposits and their formation is true or false.

(a) The charge and size of uranium ions in relation to those of other cations competing for sites in silicate minerals determine the stage in magmatic crystallization at which uranium minerals will form.

(b) Mineralized fault zones, particularly those recently discovered beneath Proterozoic erosional unconformities, provide some of the world's richest uranium ores.

(c) Uranium deposits in sandstones require higher economic cut-off grades than vein-type ores because the former are generally more expensive to mine than the latter.

(d) Roll-type uranium ores in sandstones form at an oxidation–reduction boundary where U^{6+} carried in groundwaters is converted into U^{4+} and precipitation occurs.

(e) Ionic replacement of Ca^{2+} by U^{4+} in the mineral calcite is responsible for uranium deposits in limestone.

(f) Uranium minerals may be precipitated from surface waters in the anaerobic reducing environments of lagoons and coastal swamps.

SAQ 3 (Chapter 1)   With reference to Figure 20 and 22, explain briefly in your own words why: (a) world production of uranium rose in 1980 even though the market price of uranium fell; (b) reasonably assured resources of uranium in the USA greately exceed those of Brazil, a country of comparable size; (c) so little of the Australian reasonably assured resources of uranium had been produced before 1982.

SAQ 4 (Chapter 2)   Match each of the statements and predictions A—I with the most appropriate type (or types) of geothermal energy resource (1–3).
1 Hyper-thermal resources
2 Sedimentary basins with low conductivity strata
3 Hot dry rocks

A   A non-renewable resource on the scale of human lifetimes
B   Suitable for electrical power generation
C   Requires artifical stimulation before energy can be tapped
D   Most suitable for domestic and industrial space heating
E   Producing aquifer requires impermeable seal
F   Relies on extra heat produced locally by crustal rocks rich in radioactive elements

G Relies on extra heat produced locally by recent magmatic activity

H Potentially the most prolific geothermal resource for the remainder of the twentieth century

I An inernal energy resource.

SAQ 5 (Chapter 3) The basic equation for photosynthesis reads:

$$6CO_2 + 6H_2O + 2.8 \times 10^6 J \rightarrow C_6H_{12}O_6 + 6O_2$$

i.e. 2.8×10^6 J of energy are required to form a mole of carbohydrate. How then does the calorific value of unprocessed biomass compare with that of bituminous coal ($7.9 \, kWh \, kg^{-1}$)? (Relative atomic masses are $C = 12$, $H = 1$, $O = 16$).

SAQ 6 (Chapter 3) Decide, giving reasons, whether each of the following statements about alternative energy resources is true or false.

(a) In view of the answer to SAQ 5, the char produced by pyrolysis of organic material has roughly twice the calorific value of raw biomass.

(b) For the same wind speed in each location, an aerogenerator sited on top of a hill will yield more power than one sited in shallow coastal waters.

(c) Pumped storage schemes provide one answer to the problem of using unpredictable renewable energy resources (e.g. wind and wave energies) at times of peak demand.

(d) Whereas the turbines used to exploit tidal power may be single or double direction generators, those used in oscillating water column WECS are only double direction generators.

SAQ 7 (Chapter 3)   The La Rance tidal power scheme (Chapter 3.4) can generate 520 MWh over a 4 hour period (130 MW continuously on average) during which the mean difference in water level across the barrage is 6.7 metres. If the average discharge through the turbines is $9 \times 10^3 \, m^3 s^{-1}$, what is the conversion efficiency of the tidal station during its generating period? How does this compare with the average efficiency of conventional hydropower installations? ($g = 9.8 \, m \, s^{-2}$; assume that the density of water is $10^3 \, kg \, m^{-3}$.)

SAQ 8 (Chapter 4) (a) Which of the by-products (CO_2, H_2O, CO, SO_2, nitrogen oxides and particulates) of energy production from fossil fuels is most: (i) abundant; (ii) persistent in the atmosphere; (iii) harmful to animal life; (iv) corrosive to buildings; (v) likely to induce atmospheric changes? (Note: you will need to give two answers for parts (iii) and (iv).)

(b) If use of all fossil fuels stopped tomorrow, what would happen to concentrations of these fossil fuel by-products in (i) the atmosphere, (ii) surface waters?

SAQ 9 (Chapter 4) Look back to the list of energy resources in the first paragraph of Chapter 4. Which of these resources contribute to the long-term build-up of excess heat in the atmosphere?

SAQ 10 (Chapter 5) (a) Consider the scenario whereby all the electricity used in the UK for heating purposes (2.9×10^{17} J yr^{-1}) is saved by conservation schemes coupled with locally derived renewable energy sources or CHP schemes. Assume that all this electrical energy is produced from modern coal-fired power stations at an efficiency rating of, say 29 per cent. What percentage of coal fuels (Figure 49) would be saved or be available for other uses?

(b) Given that this is coal with an averge calorific value (i.e. 7900 kWh per tonne), what does this represent in terms of coal tonnages?

SAQ 11 (Chapter 5) (a) If hydrogen is to be used as a fuel in cars of the future, then large amounts of energy will be needed to extract hydrogen from water. Why is it unrealistic to expect simple *hydrogen-burning* power stations to supply this?

(b) What other plentiful hydrogen-based power source *could* do the extraction job with energy to spare? Calculate the energy output per gram of this material (there are 6×10^{23} atoms per mole), and so deduce the corresponding amount of hydrogen that could be liberated, given an efficiency of 30 per cent.

(c) Could extraction of hydrogen from natural waters (e.g. seawater) therefore provide more than one desirable end-product?

(d) Outline briefly how technology could ultimately depend on hydrogen as an energy source for several millenia at least.

(e) What adverse by-products might there be?

SAQ answers and comments for Part I

SAQ 1 (a) This statement is true. Primary energy resources are those affecting the Earth in a natural way and winds and waves are of this character.

(b) This statement is partly false. Tidal energy is dependent on the external gravitational influence of both the Sun and the Moon. But it is correct to say that tidal bulges travel round the Earth.

(c) This statement is false. The heat flowing through the Earth's surface and produced by long-lived isotopes amounts to only about 10^{21} J yr^{-1}, whereas 2.4×10^{24} J yr^{-1} reaches the surface from the Sun.

(d) This statement is true. From the balance of incoming solar radiation and the total emission of energy in the upper atmosphere, the mean temperature of the lower atmosphere is maintained at 15°C.

(e) This statement is true. The most important heat producer is ^{235}U, and although it comprises only 0.7 per cent of natural uranium, this isotope forms the basis of nuclear fuels.

(f) This statement is false. Most of the fossil hydrocarbons are not commercially exploitable.

SAQ 2 The first important discovery in energy exploitation was that heat energy obtained from burning wood could be turned to advantage; this source was used for a long time to prepare food and to heat dwellings. And even today wood is used in this way by about half the world's population.

The second development came with the use of coal and led to the Industrial Revolution; coal replaced wood in the fabrication and manufacturing industry.

In Britain coal production reached a peak production of 285 million tonnes a year in 1913.

The third important development came about earlier this century with the mass production of cars with internal combustion engines. This caused an enormous demand for oil, and the production of oil overtook that of coal in the USA in terms of energy equivalent by the late 1940s.

SAQ 3 Coal, C; oil, B; natural gas, E; nuclear fuels, A; renewables, D.

SAQ 4 Marine bands are found all over an ancient delta and beyond it, representing a temporary marine incursion which brought marine sediments (and invertebrates) into the area of the delta. These bands must therefore represent geological processes originating outside the delta area; the most likely causes involved either regional subsidence or world-wide rises in sea-level, perhaps due to melting of polar ice-caps.

Such bands are important to the coal geologist because: (i) they can be conclusively identified by their characteristic fauna and can therefore be recognized at other localities, providing a means of correlating sequences of coal-bearing strata; (ii) they represent regional geological events occurring at specific times, and so can be used as geological datum lines between which the various seams can be grouped.

SAQ 5 The coal band is sub-divided in Figure 60 into the following six classes.
(a) Peat: decomposed fibrous plant material with high moisture content. (b) Lignite: rather soft brown coal with woody material still apparent. (c) Sub-bituminous coal: rather hard, brown coal with dull appearance developing a slight

lustre. (d) Bituminous coal: hard black material with a strongly banded appearance of alternating duller and brighter bands with a shiny lustre. (e) Anthracite: very hard, black material in which banding is no longer evident and which has a bright metallic lustre. (f) Graphite: grey form of natural carbon with a silvery lustre.

SAQ 6 (a) (i) Deposition of Carboniferous Limestone in clear seas on a seabed formed by the ancient folded and eroded Lower Palaeozoic rocks.
(ii) Deposition of sandstones, shales and coal seams in deltaic conditions.
(iii) Folding and faulting during the Variscan orogeny, followed by extensive erosion.
(iv) Deposition of sandstones and shales of Permian and Triassic ages unconformably on the eroded Carboniferous surface.
(v) Mild folding and faulting, followed by erosion to the present-day surface.
N.B. The very long period of 190M years since the deposition of the youngest Triassic rocks is unrecorded in this section. It is possible that part at least of this area was buried beneath later Jurassic and/or Cretaceous rocks which were subsequently removed by erosion, leaving no traces today.

(b) The northern limits are formed by the Wales–Brabant Ridge, and the southern limits by the folded sediments of the Variscan mountain remnants that extend east from Devon and Cornwall.

SAQ 7 The stripping ratio is the ratio of the mass of barren rock that must be removed during mining to the mass of coal extracted. The more valuable the resource then the greater the stripping ratio that can be tolerated. Anthracite fetches a considerably higher price than coal for power stations, so a higher stripping ratio can be attained on sites extracting anthracite (up to about 30:1)

than on sites producing power station coals (up to 20:1).

SAQ 8 The gradual changes include seam thinning and splitting; the sudden changes include faulting and washouts. The gradual changes would result in a deterioration in the quality of coal produced owing to the increasing proportion of dirt extracted as the coal seam thins. The sudden changes would result in the coalface being halted because the seam would be absent at the fault or washout.

SAQ 9 Power stations use the smaller grades of coal which are then pulverized to a fine powder before being burnt. Domestic consumers prefer large sizes of very clean coal and find the smaller grades unsuitable for burning in an open fire. Highly mechanized mining grinds the coal to small sizes and produces little coal suitable for the domestic consumer.

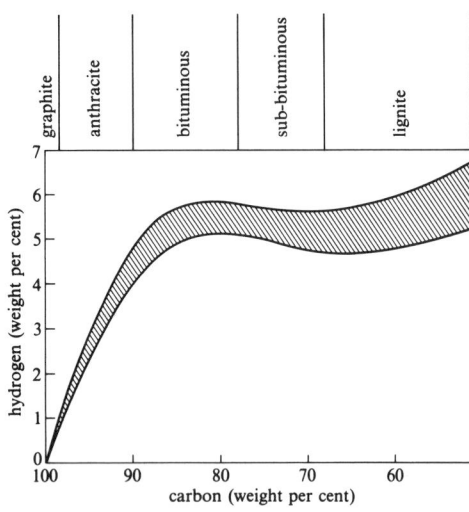

Figure 60 Answer to SAQ 5.

SAQ 10 (a) The annual production of the southern hemisphere is 310×10^6 tonnes, or 8 per cent of world production.

(b) Australia produces 136×10^6 tonnes a year, which is 44 per cent of the southern hemisphere total.

(c) The southern hemisphere total recoverable reserves, at 82×10^9 tonnes, is 16 per cent of the world total. Australia contributes 33 per cent of this. So the southern hemisphere production figure of 8 per cent is less than its contribution to the world's recoverable reserves at 16 per cent.

SAQ 11 The uppermost branch is kerogen type I, the middle branch kerogen type II and the lowest branch kerogen type III.

SAQ 12 (a) As the densest of the three substances, water will lie at the bottom of a trap and be spilt first, so will migrate furthest, i.e. to trap 3. Similarly oil underlies gas so will migrate to trap 2, leaving trap 1 filled with gas.

(b) Tar sand.

SAQ 13 Petroleum could accumulate at the following locations: 1 and 3, where the reservoir has been faulted against impermeable shale; 5, where a small anticline has formed against a fault; and a small amount at 6. At locations 2 and 4 petroleum can migrate up-dip, and there will be no oil at location 7 because the reservoir is open to the surface. However, a tar-sand deposit or an oil seepage might appear at location 7.

SAQ 14 The technique normally used to locate offshore traps is seismic reflection profiling of the rocks beneath the sea bed. Drilling is the only way to determine if there is oil in the trap. From a programme of drilling, it is possible to determine the shapes of the trap and the reservoir. It may be necessary to map the structure by drawing structure contours and isopachytes from the well data.

SAQ 15 The following data would be essential.

1 The volume of the reservoir and the porosity of the reservoir rock: from these, an estimate of the initial volume of reserve could be made.

2 The estimated percentage recovery: this depends on a number of things, including the nature of the reservoir rock and its permeability, and the quality of the crude oil.

3 The volume of oil already produced: subtraction of this from the estimate of the total recoverable volume would indicate how much recoverable oil was left in field.

SAQ 16 (a) A water well sited at locality I could be used to extract water from reservoir B and inject it into reservoir A via a borehole at 2. Similarly, gas could be extracted at 4 and injected at 3 (Figure 61). Pumping water into the existing water in reservoir A would force the oil upwards; pumping gas into the gas pocket at the top of the reservoir would force the oil downwards. It might be worth considering drilling another well to intersect the gas–oil contact to tap the oil further up in the reservoir.

(b) A is a stratigraphic trap and B is a structural trap.

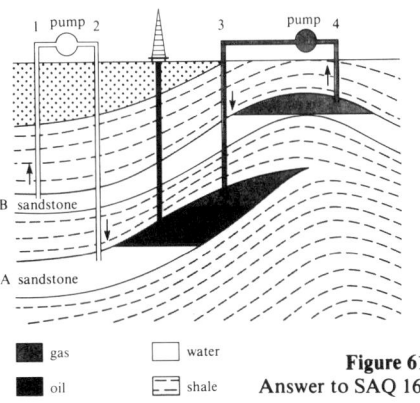

Figure 61
Answer to SAQ 16.

SAQ 17   (a)   Sandstone is the main reservoir rock.

(b)   The gas was probably derived from the coal seams of Upper Carboniferous age which underlie the region while the reservoirs are in Permian sandstones.

(c)   The main source rocks of the northern North Sea are thought to be the Kimmeridge Shales of Jurassic age. The reservoir rocks of the Forties field are Tertiary sandstones.

SAQ 18   Ideal conditions for the generation and accumulation of petroleum prevailed in the Middle East basin: high organic productivity over a very large area led to formation of huge volumes of source rocks; later folding and faulting produced numerous large traps into which the petroleum migrated.

SAQ 19   The two main factors that will affect the lifetime of the world oil reserves are (i) the percentage recovery, which will determine the volume of oil that can be obtained fromt he traps, and (ii) the rate of consumption, which will determine how long this volume will last.

SAQ answers and comments for Part II

SAQ 1 (a) in Chapter 1.1 we explained that in a burner reactor the chain reaction depends entirely on fission of ^{235}U nuclei, which capture only slow neutrons. But fission of a ^{235}U nucleus produces both fast and slow neutrons. The fast neutrons must therefore be *slowed down* so that they can be captured by other ^{235}U nuclei. Although a large number of neutrons is required to keep the chain reaction going (^{235}U constitutes only one atom in 40 of enriched uranium fuel), the reaction must be controlled. An unlimited number of neutrons would cause the fission of all the ^{235}U at once, which would be an atomic explosion! Excess neutrons above a certain number must therefore be absorbed.

(b) In burner reactors, most of the fast neutrons are moderated or absorbed — but it is not possible to remove them all. Those that remain can be captured by ^{235}U, which forms the bulk of the fuel elements in all reactors, and plutonium is formed. The breeder reactor is expressly designed to produce plutonium, because it depends on the action of fast neutrons only, which turn ^{238}U into ^{239}Pu.

SAQ 2 (a) True; see Chapter 1.2.1

(b) True; see Table 2 and Chapter 1.2.3; these are the richest large scale deposits yet discovered.

(c) False; stope mining of deep veins generally is more expensive in terms of energy and staff costs than room and pillar mining of stratiform deposits.

(d) True; see Chapter 1.2.4.

(e) False; this is the substitution that occurs in apatite, the calcium phosphate mineral containing OH^- or F^- ions which adjust numerically to compensate for the charge difference between Ca^{2+} and U^{4+}. Secondary uranium minerals occur in brecciated limestone and calcrete deposits (Chapter 1.2.6).

(f) True; these are also the environments in which coal and black shales form (Chapter 1.2.6).

SAQ 3 (a) Following the oil crisis of 1974, plans were made for massive growth in nuclear power generation and, in turn, for increasing the uranium supply by exploring in new areas and developing new mines. By 1980 it became obvious that the expansion of nuclear generation was not happening, for various reasons, resulting in over-production of uranium and a consequent fall in price during 1980. However, there is inevitably a delay in reducing the output from mines, so production continued to rise during 1980.

(b) The USA has a uranium mining industry linked to nuclear power generation that goes back many decades. It is therefore one of the countries where the most intensive efforts have been made to discover and produce uranium resources. In contrast, large scale exploration in Brazil did not commence until 1974 and the first production was not until 1981, so the figures given in Table 3 for the USA represent a much greater proportion of the resources that will ever be discovered at less than $ 130\,kg^{-1}$ than do those for Brazil. Note that there are no major geological reasons why the uranium resources of these two countries should differ.

(c) The situation in Australia is similar to that in Brazil in that there is no nuclear power industry, but differs in that exploration started much earlier, in 1947. However, there was no major uranium production in Australia until the industry as a whole underwent the expansions of the late 1970s. Several very rich, easily accessible deposits were discovered at that time.

SAQ 4 A Answers 2 and 3; hyper-thermal resources continue indefinitely so long as they are not over-produced, whereas geothermal heat is mined from sedimentary basins and hot dry rocks.

B Answers 1 and 3; steam is usually exploited for power generation in hyper-thermal volcanic areas and this, too, is the objective of investigation of hot dry rocks (Chapter 2.3). Only hot water (about 60°C, typically) is available in sedimentary basins.

C Answer 3; this refers to the explosive stimulation and hydrofracturing techniques needed to create an artificial aquifer in hot dry rocks.

D Answer 2; so far this is the only use for hot water pumped from sedimentary basins.

E Answers 1 and 2; unless there is a seal, the natural geothermal aquifers in hyper-thermal areas and sedimentary basins would 'leak'; this does not apply to hot dry rocks where the 'aquifer' is created artificially.

F Answer 3; so far, rocks of this type have been the main focus of the exploration and development of hot dry rocks (see Chapter 2.3).

G Answer 1; if you had any difficulty here, re-read Chapter 2.1.

H Answer 1; although the new technologies (2 and 3) have much to offer, they are most unlikely to overtake hyper-thermal resource output (Table 4) for the foreseeable future.

1 Answers 1–3; all forms of geothermal energy result from heat generated inside the Earth.

SAQ 5 The mass of one mole of $C_6H_{12}O_6(72 + 12 + 96g) = 180g$. On combustion, each mole will, theoretically, yield the 2.8×10^6 J that were required in its formation. So, 1 kg $(10^3 g)$ will yield:

$$\frac{10^3}{180} \times 2.8 \times 10^6 \, J = 1.56 \times 10^7 \, J$$

To convert this answer into kilowatt hours, first divide by 10^3 to give 1.56×10^4 kJ and then by 3 600 to give 4.33 kWh. So the calorific value of this type of unconverted carbohydrate biomass is 4.33 kWhkg^{-1}, just over half that of bituminous coal.

SAQ 6 (a) True; from Figure 34, char has a calorific value of about 8 kWh kg^{-1}, roughly twice the value calculated for biomass in SAQ 5 (4.33 kWh kg^{-1}).

(b) False; equation 5 shows that the power developed by an aerogenerator depends on the density of air, the blade area and the wind speed, and not on the height or the location of the windmill.

(c) True; energy can be stored by pumping water uphill to a storage reservoir during periods of slack demand and using the acquired gravitational potential energy to drive turbines during periods of peak demand.

(d) True; Figure 39 and the associated text make it clear that the power output of a tidal scheme may arise from a double or a single direction generator; often it is more economic to use the single direction type. On the other hand, half the energy from a WECS would be lost (cf. Figure 43) if only a single direction generator was used.

SAQ 7 To answer this we need equation 6, which relates power output (N) to discharge rate (Q) and working head (H):

$$N = Kg\rho QH$$

where K is the efficiency. The power output, 520MWh over 4 hours, is 130MW continuous generation ($N = 130 \times 10^6$ Js^{-1}), so

$$130 \times 10^6 = K \times 9.8 \times 10^3 \times 9 \times 10^3 \times 6.7$$

whence $K = 0.22$, or 22 per cent. This result shows that the efficiency of the tidal station is very much less than that of conventional hydropower generators (which have efficiencies exceeding 90 per cent — Chapter 3.3). As noted in Chapter 3.4, this is because of the low head available in tidal schemes.

SAQ 8 (a) (i) In absolute terms, CO_2 is the most abundant by-product of fossil fuel consumption (Table 7). Note, however, that relative to natural inputs, SO_2 is most important.

(ii) CO_2 is slowly accumulating in the atmosphere: the other by-products are absorbed biologically or washed out rapidly.

(iii) This depends on concentration; CO and the sulphur and nitrogen aerosols are all damaging to health and can kill if inhaled in sufficient quantites.

(iv) SO_2 aerosols, and to a lesser extent, those of nitrogen oxides.

(v) CO_2 is the only artificial atmospheric addition that is actually accumulating. As it absorbs long-wavelength terrestrial radiation, a significant increase in atmospheric CO_2 *could* lead to a net temperature rise over the Earth's surface.

(b) (i) Almost overnight, the concentrations of SO_2, nitrogen oxides and CO would fall rapidly to near natural levels — some might still be produced by other industrial processes but the incidence of aerosol smogs would be reduced. CO_2 would cease to accumulate, and its concentration would eventually stabilize at some new equilibrium level, probably lower than at present.

(ii) Contaminants such as SO_2 would remain in terrestrial waters rather longer, but would ultimately end up in the oceans, becoming so diluted that they would no longer be a problem.

SAQ 9 All the non-renewable resources, (B, C, D, E) give rise to excess heat because they are being consumed much faster than they can be replaced. Wood-burning (A) would also contribute to excess heat if the wood were converted at greater than replacement levels; all the remaining energy resources (F to J) have no long-term heating effect, though they might create short-term 'heat islands' in the region where they are exploited.

SAQ 10 (a) The annual saving of 2.9×10^{17} J of electrical energy is equivalent to $2.9 \times 10^{17} \times 100/29$J $= 1 \times 10^{18}$ J as coal. The annual input of coal fuels in Figure 49 is 3.47×10^{18} J; this coal is used either as solid fuel or to generate electricity. The saving in coal fuels if all heating uses of electricity were to be derived from other sources would be:

$$\frac{1 \times 10^{18}}{3.47 \times 10^{18}} \times 100 \text{ per cent} \approx 29 \text{ per cent}$$
$$\text{of coal fuels}$$

(b)
$$1 \times 10^{18} \text{ J as coal} = \frac{1 \times 10^{18}}{3.6 \times 10^6} \text{ k Wh}$$

or

$$\frac{1 \times 10^{18}}{3.6 \times 10^6 \times 7.9 \times 10^3} \text{ tonnes}$$

$$\approx 35 \text{ million tonnes}$$

SAQ 11 (a) This is a question of effi-ciency: although no specific comment on the efficiency of hydrogen-burning was made in the text, appropriate power stations are most unlikely to be 100 per cent efficient. Even hydroelectric stations are only about 90 per cent efficient and estimates for hydrogen-burning are, in fact, more like 50 per cent. So if the electrical power produced were used in the electrolytic extraction of hydrogen from water, less hydrogen would be liberated than that used to produce the power (by a factor of about 2).

(b) The answer is nuclear fusion, in which deuterium is converted into helium. For example, the fusion of two atoms of deuterium (equation 8) liberates 5.4×10^{-13} J, so the fusion of 1 g (0.5 mole) will produce

$$\frac{5.4 \times 10^{-13} \times 6 \times 10^{23}}{2 \times 2} = 8 \times 10^{10} \text{ J}$$

Even if nuclear fusion were only 30 per cent efficient, we could produce

$$\frac{8 \times 10^{10}}{1.21 \times 10^5} \times \frac{30}{100} \text{ g} \approx 2 \times 10^5 \text{ g}$$

of hydrogen from water (since, from equation 9, the burning of 1 g releases 1.21×10^5 J). So for every gram of deuterium fused, we could produce 200 000 g of hydrogen for burning.

(c) Yes, in the sense that the 200 000 g of hydrogen contains 0.015 per cent deuter-ium, i.e. 30 g of deuterium. The qualita-tive answer we expect you to have deduced is that by harnessing deuterium fusion and using the resulting power to liberate hydrogen from natural waters we would obtain not only fuel for hydrogen-burning, but also more deuterium for fusion than was fused in the first place. Oxygen gas would also be produced.

(d) Thus a balance could be struck between deuterium fusion and hydrogen-burning as energy sources, the most obvious subdivision being the use of fusion for electrical power generation and hydrogen as a propellant fuel. Thus we have a transportable energy resource that can be recycled though, of course, the deuterium fusion fuel would start to diminish in a few million years.

(e) Continual release of water vapour could lead to increased cloudiness, which would reduce the amount of solar radi-ation reaching the surface. Excess heat output could become more of a problem than today (cf. Section 4.3) and there would be some radiation hazards because fusion will produce a range of radioactive light elements. But the latter problem will be by no means as acute as that of the storage and disposal of the waste result-ing from nuclear fission.

References

This book is adapted from Block 5 of the Open University course S238 The Earth's Physical Resources.

For your information all the components of this course are listed below:–
Study Units (Printed) marketed by Open University Educational Enterprises Limited, 12 Cofferidge Close, Stony Stratford, Milton Keynes, MK11 1BY. Tel. (0908) 566744.

			ISBN Prefix 0 335
S238(1)	Resources, Economics and Geology: An Introduction	(80pp)	161553
S238(2)	Constructional and Other Bulk Materials	(72pp)	161561
S238(3)	Ore Deposits I : Origin and Distribution	(82pp)	16157X
S238(3)	Ore Deposits II : Exploration and Extraction	(84pp)	161588
S238(4)	Water Resources	(96pp)	161596
S238(5)	Energy Resources I : Fossil Fuels	(84pp)	16160X
S238(5)	Energy Resources II : Nuclear and Other Options	(80pp)	161618
S238(6)	The Future of resources : Prediction and Influence	(84pp)	161626

Videos marketed by Guild Sound and Vision Limited, 6 Royce road, Peterborough, PE1 5YB.
Tel. (0733) 315315.

S238(OOF)	Earth's Physical Resources	(50 minutes)
S238(O1V)	Copper: Resources and Reserves	(25 ")
S238(02F)	Resource Geology	(25 ")
S238(03V)	Limestones	(23 ")
S238)04V)	Clay	(22 ")
S238(05V)	Ore Genesis	(25 ")
S238(06V)	Pine Point : A Lead Zinc Deposit	(25 ")
S238(07V)	Pine Point : Origin and Exploration	(25 ")
S238(08V)	Pine Point : Ore to Metal	(23 ")
S238(09RV)	Water for a City	(24 ")
S238(10F)	Water for Jordan	(25 ")
S238(11V)	Energy Resources – Coal	(25 ")
S238(12V)	Energy Resources – Petroleum	(25 ")
S238(13F)	Energy from the Crust : Uranium	(24 ")
S238(14F)	Energy : The Alternatives	(25 ")
S238(15RV)	Oil : Finds for the Future	(25 ")
S238(16V)	Minerals : Finds for the Future	(24 ")

Audio-Visual Pack (Boxed) marketed by Open University Educational Enterprises Limited, 12 Cofferidge Close, Stony Stratford, Milton Keynes, MK11 1BY.
Tel. (0908) 566744.

S238(A/V) 3 Audiocassettes together with accompanyingprinted booklets and colour plate booklets on the topics:–

Minerals and Rocks
The Postcard Geological Map
Clay Minerals
Strike, Dip and Cross-sections
Ores and Ore Minerals
Remote Sensing
The Hydrological cycle
Water Transfer
Maturation of Coal and Petroleum
Solar and Biomass Energy
Prediction and Influence

Appendix

Energy Units and conversion factors

1 foot = 0.305 m
1 mile = 1609 m
1 hectare = 10^4 m^2

1 tonne = 10^3 kilogram (kg)
1 kilogram = 2.205 pounds (1b)

1 litre = 10^{-3} m^3 (10^3 cm^3)
1 cubic metre = 35.3 cubic feet (ft^3)

1 barrel oil (bbl) = 35 Imperial gallons
= 42 US gallons
= 0.159 m^3
1 tonne = 6.5–7.5 bbl depending on density

1 calorie = 4.19 joules
1 kilowatt hour (kWh) = 3.6×10^6 joules
1 therm = 29.3 kWh
1 kilowatt = 10^3 joules per second
= 3.15×10^{10} joules per year

parts per million (p.p.m.) = milligrams per litre (mg 1^{-1})

Prefix	Symbol	Power of 10
giga	G	10^9
mega	M	10^6
kilo	k	10^3
hecto	h	10^2
centi	c	10^{-2}
milli	m	10^{-3}
micro	μ	10^{-6}
nano	n	10^{-9}
pico	ρ	10^{-12}

10^3 is one thousand
10^6 is one million
19^9 is one billion

Index

Acceleration due to gravity   158
Accessory mineral   107
Acid, rain   174
Active continental margin   6
Advance face   40, 43
Advanced gas-cooled reactor (AGR)   99
Aerogenerator   156, 157
Aerosol   174
Algae   56
Alpha particle   95
Alternative energy resources   144
Amplitude   166
Anaerobic (anoxic) conditions   7, 13, 20, 55, 150
Andesite   130
Anhydrite   67
Anthracite   20, 47
Anticline   30
Anticlinal trap   65, 133
Aquifer   64, 130, 132
Ash   23
Asphalt   55
Atmospheric carbon dioxide   171,174
Atmospheric dust   173
Atmospheric nitrogen oxide   174
Atmospheric sulphur dioxide   173, 174
Atmospheric water vapour   172, 174
Avogadro constant   96

Back-pressure power plant   134, 135
Bench   39
Beta decay   96
Biochemical decomposition   19
Biogas   149
Biomass energy conversion   6, 149, 153
Bituminous coal   20
Boghead coal   22
Burner reactor   97

Calcrete   116
Calorific value   23, 150
Channel coal   22
Capilliary action   116
Cap-rock   67
Carbon cycle   171
Cartel   10
Catagenesis   57
Catalytic process   174
Channel-fill deposit   42
Chain reaction   96
Char   150
Clastic sediments   24
Closure   66
Coal-face   40
Coalification   20
Coal seam   16, 39
Coal washery   44
Coke   23, 47
Coking coal   23, 47
Combination trap   65, 69
Combined   heat   &   power   (CHP) station   141, 154

Concealed coalfield   31
Concretion   24
Condensation power plant   134, 135
Conditional resources   49, 122
Confined aquifer   74
Conglomerate   106, 115
Constructive plate margin   130
Continental crust   109
Continent-continent collision   86
Control rod   97
Core barrel   32
Core logging   33
Cracking   57
Craton   15, 109
Crude oil   55
Cut-off grade   105

Deltaic environment   18
Depleted uranium   101
Destructive plate margin   130
Diagenesis   24, 57
Dirt band   19
Disseminated magmatic deposit   105, 108
Distributory   18
District heating   137
Dolomite   80
Dome   65
Doubling time   12
Dragline   39
Drape   66
Dry gas   59

Eh-pH diagram   107
Electronegativity   107
Endothermic reaction   6, 188
End-use energy   181
Energy capital   1
Energy conservation   183
Energy flow in the UK   181, 182
Energy gap   13
Energy income   1
Enhanced recovery   74
Enriched uranium   97
Estimated additional resources   123, 124
Ethanol   150
Evaporite   27
Excess heat   179
Exinite   22, 61
Exothermic reaction   6, 188
Exposed coalfield   31
Exploration   32, 35, 72
External energy source   2

Fast breeder reactor   97, 99
Fast neutron   96
Fault breccia   110
Fault trap   67
Fireclay   17
Fission   95, 176
Fission energy   96
Flat plate collector   146
Flotation method   43

Fluidized bed combustion   47
Flux   17
Focussing collector   146
Fossil fuels   6, 7, 13
Fuel cell   188

Ganister   17
Gas field   22, 63, 77
Geophysical logging   33
Geothermal gradient   42, 58, 130
Geothermal heat   130
Glycolysis   150
Graben   79
Granodiorite   105
Graphite   20
Greenhouse effect   4, 171
Gypsum   67

Half-live   176
Halite   67
Heat flow   130
Heat load   138
Heat pump   139
HDR geothermal resources   139
HBMIC   22
Hydrocarbons   7
Hydro-fracturing   141
Hydrogen economy   189
Hydrogenation   48
Hydrological cycle   2, 158
Hydropower   158, 160
Hydrothermal convection   132
Hydrothermal vein deposit   13, 105, 110
Hyper-thermal area   130
Hypothetical resources   122

Illite   24
Inertinite   22, 59
Intermediate and high-level
   radioactive wastes   177
Internal energy source   2, 130
Ionic charge   106
Ionic size (radius)   106
In-seam seismics   43
Isopachyte   43, 73

Jig   44
Joule   1

Kaolinite   24
Kerogen   57
Kilowatt hour   1
Kinetic energy   2, 144

Lead pollution   175
Lead time   160
Lignite   20
Liquifaction   88
Lithification   64
Longwall mining   40
Low-level radioactive wastes   177

Marceral   22
Magma   106
Magnox reactor   99
Marine band   17, 33
Marker horizon   19
Maturation   20, 22, 57
Metagenesis   57
Moderator   97

Natural gas   22, 55, 69
Neutron   95
Non-renewable resources   1, 137
Nuclear fuel   6, 96, 102
Nuclear fusion   2, 188
Nuclei   95

Ocean-ridge   6
Oil-field   63, 76
Oil sand   70
Oil shale   71
Opencast mining   39, 45
Openpit mining   111, 119
Organisation of Petroleum
   Exporting countries (O.P.E.C.)   10,
Orogenesis   22
Oscillating water column
   converter   167
Overburden   39
Oxidation potential (Eh)   107, 112

Palaeomagnetism   26
Peat   16, 20
Pegmatite   105, 108
Permeability   64, 110
Petroleum   55
Petroleum fractions   55
Photosynthesis   2, 6, 145, 149
Phytoplankton   6, 56
Piezometric pressure   138
Piezometric surface   74
Pitchblende   110
Pitch lakes   70
Place value   13
Placer deposit   106
Polyanions   107
Pore space   64
Porosity   1, 116
Pressurized water reactor
   (PWR)   99
Primary energy resources   1, 130, 144
Primary migration   63
Primary recovery   74
Primary uranium deposit   106, 118
Principle of Uniformitarianism   18
Projected energy demand   12
Pulverized fuel   47
Pumped storage scheme   159
Pyrolysis   150

Radioactive decay   6, 95
Radioactive waste disposal   176

Rank   20
Raw coal   44
Reaction rate   97
Reactor Core   97
Reasonably assured resources   122, 124
Recovery factor   43
Recharge (of an aquifer)   133
Redox reaction   112
Reef trap   69
Reflectance   23
Refuse derived fuels (RDF)   153
Relative atomic mass   96
Renewable resources   1, 132, 187
Reprocessing of nuclear fuels   104
Reserves   37, 49, 53, 84, 122
Reservoir rock   63, 77
Retreat face   40, 43
Rift-drift sequence   79
Rock cycle   109
Roll-type uranium ores   112, 113
Room and pillar mining   117

Salt plug   67
Sapropelic coal   22
Screen   44
Seal   64, 132
Seatearth   17
Secondary energy resource   1
Secondary migration   63
Secondary porosity   69
Secondary recovery   74
Secondary uranium deposit   106, 118
Sedimentary basin   63, 112
Seismic reflection profiling   34, 43, 72
Semi-thermal area   130
Shearer   40
Slow neutron   95
Solar cell   148
Solar energy   144, 151
Source rock   56, 79
Spill-point   66
Steam flashing   132
Stockwork   105
Stope mining   117

Strata volume method   87
Stratigraphical trap   65
Stripping ratio   40
Structure contour   72
Structural trap   65, 111
Sub-bituminous coal   20
Submarine fan   80
Syenite   109
Syncline   30
Synthetic crude oil   71
Synthetic natural gas (SNG)   48

Tar sands   70
Tectonic plates   6
Thermal conductivity   130
Thermal oxide re-processing
   plant (THORP)   105
Threshold (of petroleum
   generation)   62
Tidal barrage   162, 165
Tidal power generation   5, 162
Trace element   24
Trap   64, 87

Unconformity   31, 105
Unconformity trap   69, 111
Uranium ore deposit   105

Variscan orogeny   30, 110
Vein-unconformity deposit   105, 110
Vitrinite   22, 59
Volatile matter   21

Washout   43
Water table   16
Wave energy converter
   system (WECS)   167
Wedge-edge trap   69
Wet gas   59
Wind energy   155, 156
Working head   158

Yellowcake   119, 120